T0295806

# Data Science for Sensory and Consumer Scientists

*Data Science for Sensory and Consumer Scientists* is a comprehensive textbook that provides a practical guide to using data science in the field of sensory and consumer science through real-world applications. It covers key topics including data manipulation, preparation, visualization, and analysis, as well as automated reporting, machine learning, text analysis, and dashboard creation. Written by leading experts in the field, this book is an essential resource for anyone looking to master the tools and techniques of data science and apply them to the study of consumer behavior and sensory-led product development. Whether you are a seasoned professional or a student just starting out, this book is the ideal guide to using data science to drive insights and inform decision-making in the sensory and consumer sciences.

Key Features:

- Elucidation of data scientific workflow.
- Introduction to reproducible research.
- In-depth coverage of data-scientific topics germane to sensory and consumer science.
- Examples based in industrial practice used throughout the book

## Chapman & Hall/CRC Data Science Series

Reflecting the interdisciplinary nature of the field, this book series brings together researchers, practitioners, and instructors from statistics, computer science, machine learning, and analytics. The series will publish cutting-edge research, industry applications, and textbooks in data science.

The inclusion of concrete examples, applications, and methods is highly encouraged. The scope of the series includes titles in the areas of machine learning, pattern recognition, predictive analytics, business analytics, Big Data, visualization, programming, software, learning analytics, data wrangling, interactive graphics, and reproducible research.

Published Titles

*Data Science: An Introduction*
Tiffany-Anne Timbers, Trevor Campbell and Melissa Lee

*Tree-Based Methods: A Practical Introduction with Applications in R*
Brandon M. Greenwell

*Urban Informatics: Using Big Data to Understand and Serve Communities*
Daniel T. O'Brien

*Introduction to Environmental Data Science*
Jerry Douglas Davis

*Hands-On Data Science for Librarians*
Sarah Lin and Dorris Scott

*Geographic Data Science with R: Visualizing and Analyzing Environmental Change*
Michael C. Wimberly

*Practitioner's Guide to Data Science*
Hui Lin and Ming Li

*Data Science and Analytics Strategy: An Emergent Design Approach*
Kailash Awati and Alexander Scriven

*Telling Stories with Data: With Applications in R*
Rohan Alexander

*Data Science for Sensory and Consumer Scientists*
Thierry Worch, Julien Delarue, Vanessa Rios de Souza and John Ennis

For more information about this series, please visit:
https://www.routledge.com/Chapman–HallCRC-Data-Science-Series/book-series/CHDSS

# Data Science for Sensory and Consumer Scientists

Thierry Worch
Julien Delarue
Vanessa Rios de Souza
John Ennis

CRC Press is an imprint of the
Taylor & Francis Group, an **informa** business
A CHAPMAN & HALL BOOK

Designed cover image: © Shutterstock, ID 1130063375, anttoniart

First edition published 2024
by CRC Press
6000 Broken Sound Parkway NW, Suite 300, Boca Raton, FL 33487-2742

and by CRC Press
2 Park Square, Milton Park, Abingdon, Oxon, OX14 4RN

*CRC Press is an imprint of Taylor & Francis Group, LLC*

© 2024 Taylor & Francis Group, LLC

Reasonable efforts have been made to publish reliable data and information, but the author and publisher cannot assume responsibility for the validity of all materials or the consequences of their use. The authors and publishers have attempted to trace the copyright holders of all material reproduced in this publication and apologize to copyright holders if permission to publish in this form has not been obtained. If any copyright material has not been acknowledged please write and let us know so we may rectify in any future reprint.

Except as permitted under U.S. Copyright Law, no part of this book may be reprinted, reproduced, transmitted, or utilized in any form by any electronic, mechanical, or other means, now known or hereafter invented, including photocopying, microfilming, and recording, or in any information storage or retrieval system, without written permission from the publishers.

For permission to photocopy or use material electronically from this work, access www.copyright.com or contact the Copyright Clearance Center, Inc. (CCC), 222 Rosewood Drive, Danvers, MA 01923, 978-750-8400. For works that are not available on CCC please contact mpkbookspermissions@tandf.co.uk

*Trademark notice:* Product or corporate names may be trademarks or registered trademarks and are used only for identification and explanation without intent to infringe.

*Library of Congress Cataloging-in-Publication Data*
Names: Worch, Thierry, author. | Delarue, Julien, author. | De Souza, Vanessa Rios, author. | Ennis, John M., author.
Title: Data science for sensory and consumer scientists / Thierry Worch, Julien Delarue, Vanessa Rios De Souza and John Ennis.
Description: Boca Raton : CRC Press, 2024. | Series: Chapman & Hall/CRC data science series | Includes bibliographical references and index.
Identifiers: LCCN 2023006718 (print) | LCCN 2023006719 (ebook) | ISBN 9780367862879 (hardback) | ISBN 9781032384962 (paperback) | ISBN 9781003028611 (ebook)
Subjects: LCSH: Statistics--Data processing. | Food--Sensory evaluation--Data processing. | Consumers--Research--Data processing. | R (Computer program language)
Classification: LCC QA276.4 .W67 2024 (print) | LCC QA276.4 (ebook) | DDC 519.50285--dc23/eng20230715
LC record available at https://lccn.loc.gov/2023006718
LC ebook record available at https://lccn.loc.gov/2023006719

ISBN: 978-0-367-86287-9 (hbk)
ISBN: 978-1-032-38496-2 (pbk)
ISBN: 978-1-003-02861-1 (ebk)

DOI: 10.1201/9781003028611

Typeset in CMR10
by SPi Technologies India Pvt Ltd (Straive)

*To Luca,*

*To Bella*

# Contents

# Preface

## Who Should Read This Book?

This book is for practitioners and students of sensory and consumer science who want to participate in the emerging field of computational sensory science. This book assumes little to no coding experience. Some statistical experience will be helpful to understand the examples discussed.

## How Is This Book Structured

It is important to start by saying that the aim of this book is neither to explain in depth what the different sensory and consumer methods are nor to explain how the data gathered from these methods should be analyzed. For such topics, other excellent books including Lê and Worch (2018), Lawless and Heymann (2010), Civille and Carr (2015), Stone et al., (2020) for example are available.

Instead, the aim is to explain the workflow sensory and consumer scientists can adopt to become more efficient and to push their analyses further. The workflow proposed includes many steps including:

- Setting up the test both from an experimental design and an analysis perspective (e.g. setting up projects, collaboration tools, etc.);
- Data collection and data processing through data cleaning, data manipulation and transformation, and data analysis;
- Communication of the results (e.g. visualization, reporting, communication).

## How to Use This Book

This book is meant to be interactive, with the reader ideally typing and running all of the code presented in the book. Computer languages are like human languages in that they need to be practiced to be learned, so we highly recommend the reader actually typing all of the code from these parts, running it, and verifying they obtain the results shown in the book. To help with this practice, we have created a special GitHub repository[1] that contains folders called `code` and `data`. Please see Chapter 2 for guidance on how to get started with R and GitHub. In the `code` folder, we have included the code as presented in the book, while in the `data` folder we have provided the data sets used throughout the book.[2] Our advice is to download the files and store them on your local machine in a folder named *data*. This solution was adopted throughout the book.

## Acknowledgments

First and foremost, we all thank our partners for their continual support. Arkadi Avanesyan, Bartosz Smulski and Tian Yu all contributed to the content in various important ways. We also would like to Dr. Jacob Lahne for kindly sharing the data set used in Chapter 13. Last but not least, we'd like to address a warm thank you to our respective families for their patience, support, and constant encouragement. This book would have never seen the light without you.

[1] https://github.com/aigorahub/data_science_for_sensory

[2] You may need to load these libraries before the code will run. Please see Appendix 2 for more information on this topic as well.

# About the Authors

**Thierry Worch** is a data enthusiast who engineers Simple, Meaningful, Automated, Reproducible, Trustworthy solutions to Sensory and Consumer challenges at FrieslandCampina. Back in 2009, he started as a project manager at OP&P Product Research (Utrecht, The Netherlands) before completing his PhD in 2012 on "the Ideal Profile Analysis: from the validation to the statistical analysis of ideal profile data" both in collaboration with Pieter Punter (OP&P) and Sébastien Lê and Jérôme Pagès (Agrocampus-Ouest). By the end of 2012, he joined Qi Statistics as a consultant where he ran clients' projects, participated in reseach projects, and gave diverse trainings. During that time, he also continued a long collaboration with Logic8 (EyeQuestion Software) where he produced all the R-routines included in the EyeOpenR software. By the end of 2019, he joined the Global Sensory Department of Friesland Campina, Wageningen, The Netherlands, where he provides his expertise in Sensory and Consumer Methods, Sensometrics, and Data Science. Besides publishing various papers related to Sensometrics and contributing to many books, he is also the co-author with Sébastien Lê of the book entitled *Analyzing sensory data with R* (CRC Press).

**Julien Delarue** is an associate professor at the University of California Davis. He received his PhD in Food Science at AgroParisTech, and his research focuses on methods to measure sensory perception and preferences and on their effective use in food design. He explores the role of context in hedonic measures using immersive environments and digital technologies. He also works to develop and validate rapid and flexible descriptive analysis methods with application to new product development and consumer research. Formerly a professor at AgroParisTech, France, in the food science and technology joint research unit with INRAE and Université Paris-Saclay, he has served as the Chair of the French Society for Sensory Analysis (SFAS) and of the European Sensory Science Society (E3S).

**Vanessa Rios de Souza** is a Sr. Computational Sensory Science Consultant at Aigora, a company whose mission is to empower sensory and consumer science teams to implement artificial intelligence. Her role involves leading clients' projects related to Process Automation, Knowledge Management, New Technologies, and Computational Analytics. Vanessa holds a PhD degree in Food Science and has over 10 years of experience in R&D and consumer and sensory research across multiple food product categories and functions, with a strong background in food science and food processing. She has extensive experience in industrial, academic, and research settings.

**John Ennis** is the co-founder of Aigora and a world-renowned authority in the use of artificial intelligence within sensory and consumer science. He is a PhD mathematician who conducted his postdoctoral studies in computational neuroscience and who has more than a dozen years of experience as a sensory and consumer science consultant.

# 1

# *Bienvenue!*

## 1.1 Why Data Science for Sensory and Consumer Science?

Located at the crossroads of biology, social science, and design, sensory and consumer science (SCS) is definitely the tastiest of all sciences. On the menu is a wide diversity of products and experiences that sensory and consumer scientists approach through the lens of human senses. Thanks to a wide set of refined methods, they have access to rich, complex, and mouthwatering data. Delightedly, data science empowers them and leads them to explore new flavorful territories.

### 1.1.1 Core Principles in Sensory and Consumer Science

Sensory and consumer science is considered as a pillar of food science and technology and is essential to product development, quality control, and market research. Most scientific and methodological advances in the field are applied to food. This book makes no exception as we chose a cookie formulation data set as a main thread. However, SCS widely applies to many other consumer goods, so are the contents of this book and the principles set out below.

#### *Measuring and Analyzing Human Responses*

Sensory and consumer science aims at measuring and understanding consumers' sensory perceptions as well as the judgment, emotions, and behaviors that may arise from these perceptions. SCS is thus primarily a science of measurement, although a very particular one that uses human beings and their senses as measuring instruments. In other words, sensory and consumer researchers measure and analyze human responses.

To this end, SCS relies essentially on sensory evaluation which comprises a set of techniques that mostly derive from psychophysics and behavioral research. It uses psychological models to help separate signal from noise in collected data (Lee and O'Mahony, 2004; Ennis, 2016). Besides, sensory

DOI: 10.1201/9781003028611-1

evaluation has developed its own methodological framework that includes most refined techniques for the accurate measurement of product sensory properties while minimizing the potentially biasing effects of brand identity and the influence of other external information on consumer perception (Lawless and Heymann, 2010).

A detailed description of sensory methods is beyond the scope of this book and many textbooks on sensory evaluation methods are available to readers seeking more information. However, just to give a brief overview, it is worth remembering that sensory methods can be roughly divided into three categories, each of them bearing many variants:

- Discrimination tests that aim at detecting subtle differences between products.
- Descriptive analysis (DA), also referred to as "sensory profiling", aims at providing both qualitative and quantitative information about products' sensory properties.
- Affective tests. This category includes hedonic tests that aim at measuring consumers' liking for the tested products or their preferences among a product set.

Each test category generates its own type of data and related statistical questions in relation to the objectives of the study. Typically, data from difference tests with forced-choice procedures (e.g. triangle test, duo-trio, 2-AFC, etc.) consist of a series of binary answers (correct/failed) depending on whether judges successfully picked the odd sample(s) among a set of three or more samples.[1] These data are used to determine whether the number of correct choices is above the level expected by chance (see O'Mahony and Rousseau, 2003, for an overview of these methods, the related theories, and experimental factors).

Conventional descriptive analysis data consist of intensity scores given by each panelist to evaluated samples on a series of sensory attributes, hence resulting in a product x attribute x panelist data set (Figure 1.1). Note that depending on the DA method, quantifying means other than intensity ratings can be used (ranks, frequency, counts, etc.). Most frequently, each panelist evaluates all the samples in the product set. However, the use of a balanced incomplete design can also be found when the experimenters aim to limit the number of samples evaluated by each subject.

Eventually, data sets from hedonic tests consist of hedonic scores (i.e. degrees of liking or preference ranks) given by each interviewed consumer to a series of products (Figure 1.2). As in the case of DA, each consumer usually evaluates all the samples in the product set, but balanced incomplete

[1] Other procedures like the *different from control* test or the *degree of difference* test generate rating data.

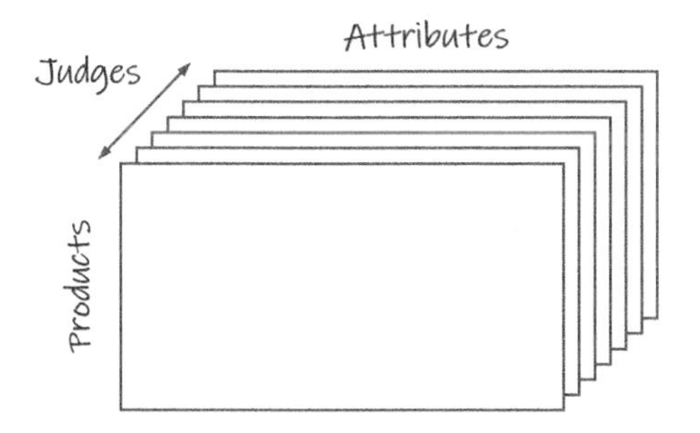


**FIGURE 1.1**
Typical structure of a descriptive analysis data set.

designs are sometimes used too. In addition, some companies favor pure monadic evaluation of products (i.e. between-subject design or independent group design), which obviously result in unrelated sample data sets.

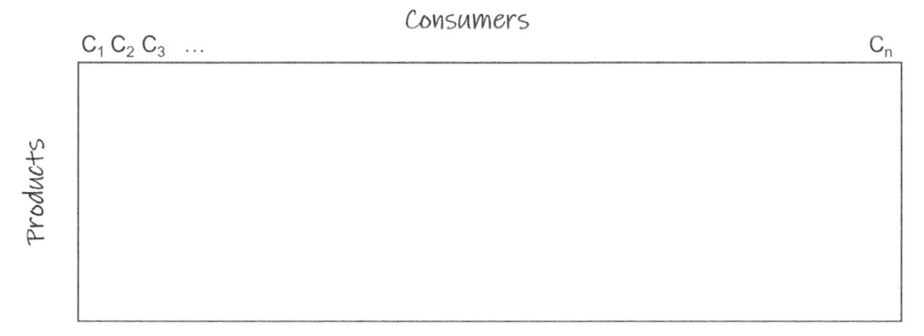


**FIGURE 1.2**
Two-way hedonic data from a consumer test where "n" consumers have evaluated a series of products.

Sensory and consumer researchers also borrow methods from other fields, in particular from sociology and experimental psychology. As a result, it is now frequent to collect textual sensory data from open comments and qualitative interviews, or sensory distances or co-occurrences from projective and sorting tasks. Definitely a multidisciplinary area, SCS develops in many directions and reaches disciplines that range from genetics and physiology to social marketing, behavioral economics, and computational neuroscience. This has diversified the types of data sensory and consumer scientists must deal with. As in many scientific fields, the development of sophisticated statistical techniques and access to powerful data analysis tools have played an important role in the evolution of sensory and consumer science. Statisticians and data analysts in SCS have developed their own field of research, coined

Sensometrics (Schlich, 1993; Brockhoff, 2011; Qannari, 2017). Now then, what makes sensory and consumer science special? And how does it influence the way sensory and consumer data are handled?

### *Dealing with Human Diversity*

Sensory evaluation attempts to isolate the sensory properties of foods and provides important and useful information about these properties to product developers, food scientists, and managers (Lawless and Heymann, 2010). However, one should bear in mind that these "sensory properties" actually result from the interaction between the object (the food) and the perceiver of that object (the consumer). In fact, we may very well consider the true object of evaluation in SCS to be mental representations. They are nonetheless very concrete and directly impact behaviors, health, and economic decisions (Kahneman and Tversky, 2000). A direct consequence of this is that sensory data depend both on the product to be evaluated and on the subjects who evaluate the product. Because people are different, individual sensory data are expected to differ accordingly. In its core principle, SCS recognizes the diversity of human beings, biologically, socially, and culturally speaking, not to mention the fact that each individual has their own personal history and experience with products. In short, people perceive things differently and like different things. For this reason, SCS only relies on groups of people (i.e. a panel of judges, a sample of consumers) and never on a single individual's response. Yet, sensory and consumer scientists usually collect individual data and analyze them at a refined level of granularity (individual, subgroups) before considering larger groups (specific populations or marketing targets).

This said, sensory and consumer studies must lead to operational recommendations. They are used to make informed decisions on product development, to launch a new product, and sometimes to help define food and health policies. Data science can precisely help sensory and consumer scientists to reach those objectives while taking diversity into account.

For measures of sensory description, sensory and consumer scientists can manage the diversity of human responses to a certain extent by training panels to use a consensual vocabulary, by aligning evaluated concepts, and by calibrating the quantification of evaluations on scales (Bleibaum, 2020). However, this won't eliminate interindividual differences in sensitivity, simply because we are genetically different, on top of differences due to age, physiological state, illness, etc. Nowadays, as the field becomes more and more consumer-oriented, it becomes clear that the use of several subjects in a panel cannot be assimilated to a mere repetition of individual measurements. Accordingly, sensory methods have been developed to allow panelists to better express their own perceptions and to get a more accurate picture of how people perceive products (Varela and Ares, 2012). These methods yield richer and more complex data that require more advanced analysis techniques to extract relevant and actionable information. Historically, one the first methodological innovations in this direction has been the use of free choice profiling combined with Generalized Procrustes Analysis (Williams and Langron, 1984). Since then, sensory and data analysis methods have multiplied greatly (Delarue and Lawlor, 2022). Naturally, data science has become even more crucial to accompany this evolution.

As regards hedonic tests (liking, acceptance, preference...), the measurements are in essence subjective and participants of such tests are by definition "untrained" consumers. A constant outcome of these tests is to find important interindividual differences and it is very common to find consumers who have opposite sensory preference patterns. Clustering and segmentation techniques are thus routinely applied to consumer data. One difficulty though is to link these differences in preferences to other consumer variables, should they be sociodemographic, psychographic, or related to usage and attitudes. Most often, one set of variables (e.g. demographics) is not enough to fully explain preference patterns. In saturated and ever changing markets, however, being able to understand individual needs and preferences is critical should one intend to develop customized products. This makes the understanding of consumer segments even more important. Nowadays, these segments go far beyond sensory preferences and must take into account variables that touch environmental awareness and sustainability dimensions.

### *Specificities of Data Handled in Sensory and Consumer Science*

Sensory and consumer data are usually of relatively small size. Indeed, we often deal with a number of subjects ranging between a dozen (for trained panels) and few hundreds (for consumer hedonic tests). Of course, when multiplied by the number of samples being evaluated by each subject, we would get a much larger numbers of observations, but this will still be relatively modest compared with so-called big data generated everywhere online. The same goes with the number of variables in sensory data sets. Sensory descriptive analysis, for example, typically relies on 10–50 attributes, which could be seen as a lot but is in fact much less than in other fields producing experimental data with thousands of variables like chemometrics, genomics, etc.

This being said, it must be stressed that sensory and consumer data are very diverse. Indeed, the need to understand perceptions and preferences often leads sensory and consumer scientists to deal with multiple data sets, each possibly comprising various data types (Figure 1.3). Most sensory techniques yield quantitative (e.g. intensity, similarity, and hedonic) data collected from rating scales or ranking tasks, but other methods would provide inter-product distances (e.g. napping), co-occurrences (e.g. free sorting), citation frequencies (e.g. CATA), or texts (e.g. open-ended comments, natural speech). Besides, agreement scores from Likert scales would often be used when sensory studies are combined with usage and attitude surveys or psychometric questionnaires. To add richer information, but more complexity to this picture, experimenters are sometimes interested in the temporal dimension of sensory measurements (by the means of methods like TI, TDS, TCATA) or may simply aim to measure reaction times (e.g. Implicit Association Test).

Eventually, different types of data can result from the same task. For example, this would typically be the case for *free JAR* that yields both categorization data with hedonic valence and textual data (Luc et al., 2022a). With the development of all sorts of media and data collection means, such patterns will surely become even more frequent.

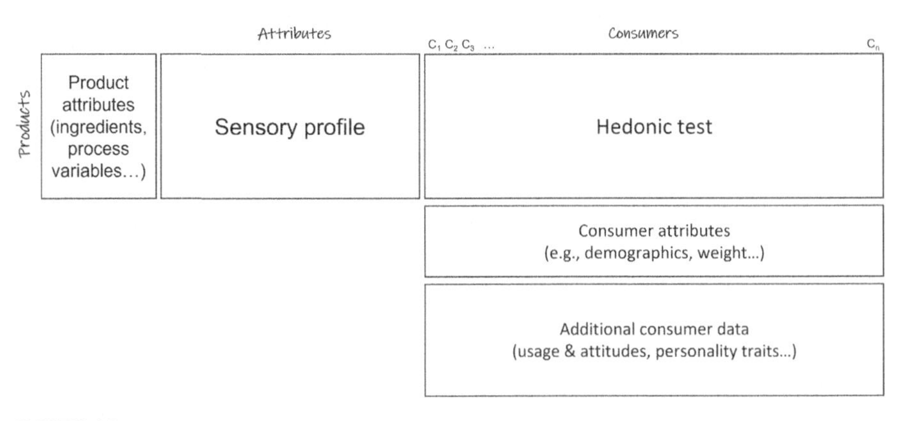


**FIGURE 1.3**
Sensory and consumer science studies often yield multiple data sets that may all relate to each other.

As could be expected, sensory and consumer studies are often multifaceted and collected data may all relate to each other when they apply to the same product set and/or to the same consumers. Such links between data sets are usually sought because they allow uncovering consumers' motivations and their drivers of preferences, thanks to modeling techniques (e.g. preference mapping, PLS regression), segmentation analyses (e.g. latent variables clustering), and machine learning. As a prerequisite to the application of any of these techniques though, it is critical to understand how these data are structured and to properly handle them in a reliable and efficient manner. Many examples of such data manipulation are given throughout this book and specific guidance is given in Chapter 4.

Last, it is worth mentioning that sensory and consumer data are intrinsically subjective. This is of course a good thing because the goal of any sensory study is to capture subjects' point of view. However, it could make some of the usual data quality criteria useless. This is specially true for hedonic data, for which repeatability and reference values could be questionable notions (Köster, 2003;

Köster et al., 2003). Sensory and consumer scientists may nonetheless rely on techniques allowing them to evaluate the degree of consensus of their panel or tools like jackknife and bootstrap to evaluate the robustness of their data.

### 1.1.2 Computational Sensory Science

We can make an analogy of the future (or maybe the present already) of the sensory and consumer science field with other areas that advanced into the computational field, such as computational neuroscience and computational biology. A quick search in Wikipedia on the definition of those fields and a little about on how those areas evolved or how the term "computational" was introduced will make you realize that is the same path as the consumer and sensory field is moving along.

- **Computational neuroscience**: "*is a branch of neuroscience which employs mathematical models, computer simulations, theoretical analysis and abstractions of the brain to understand the principles that govern the development, structure, physiology and cognitive abilities of the nervous system. The term 'computational neuroscience' was introduced to provide a summary of the current status of a field which was referred to by a variety of names, such as neural modeling, brain theory and neural networks.*" https://en.wikipedia.org/wiki/Computational_neuroscience
- **Computational biology**: is a branch of biology that "*involves the development and application of data-analytical and theoretical methods, mathematical modelling and computational simulation techniques to the study of biological, ecological, behavioral, and social systems. Computational biology, which includes many aspects of bioinformatics and much more, is the science of using biological data to develop algorithms or models in order to understand biological systems and relationships.*" https://en.wikipedia.org/wiki/Statistical_model

The sensory and consumer science field, although not officially named with the term "computational", is already expanded in this field. The way consumer and sensory data is explored today is extremely advanced and went way beyond the simple statistical analysis performed a few years ago using the data collected from standard consumers or trained panel studies. Nowadays, sensory is getting into the big data field by organizing and putting together years of historical data into a database to answer future business questions and extract meaningful information.

Advances in digital technologies such as the integration of biometrics to assess consumers' physiological and emotional responses, incorporation of virtual, augmented, and mixed reality, and even the use of sensor technologies (*electronic noses and tongues*) for sensory analysis are already widely used in the field. Additionally, data are being collected from different sources, such as social media. Those advanced technologies and complex data being extracted require much more advanced tools and computer capabilities to analyze and get meaningful information.

Rapid data acquisition, allied with the need for flexible, customized, and fast result interpretation, is opening a huge way for automation. The urge to deep explore, segment products/consumers, and discover new or hidden patterns and relationships to get the most valuable insights from the data sets is also nurturing the implementation of Artificial Intelligence, particularly Machine Learning.

At this point, we hope to have motivated you even more about the importance of data science for practitioners and students of sensory and consumer science who want to participate in the emerging field of computational sensory science.

Let's get started?

# 2

# *Getting Started*

## 2.1 Introduction to R

### 2.1.1 What Is R?

First released in 1995, R is an open-source programming language and software environment that is widely used for statistical analyses, graphical representations, and reporting. R was created by Ross Ihaka and Robert Gentleman at the University of Auckland, New Zealand, and is currently developed by the R Development Core Team (R Core Team, 2022).

R is a scripting language (not a compiled language) that runs the lines of code or commands one by one, in order. It is one of the most popular languages used by statisticians, data analysts, researchers, marketers, etc. to retrieve, clean, analyze, visualize, and represent data. By the time this book is being written, it is among the most popular programming languages in the world, including the sensory and consumer science field.

---

Do you know why R is called as such? It seems that the name *R* has two potential origins: It is the first letter of both Ihaka's and Gentleman's first names, but also it is a play on the name of Bell Labs Software called S it originates from (a lot of code that runs in S also run in R).

---

### 2.1.2 Why Learn R (or Any Programming Language)?

There are several reasons why you should learn R or any programming language for that matter.

First, it gives the user a lot of **control**. Compared to other statistical software, which can be seen as a *black box* (you do not have necessarily access to the code that runs behind the scene) and are restricted to the features their developers provide, R allows you to see what is happening at each step of your analysis (you can print the code that runs behind each function to ensure that it does what you are expecting...) and allows exploring any type of analysis. This means that users are fully in control and are only limited by

DOI: 10.1201/9781003028611-2

their imagination (and maybe their program skills?). A direct advantage of this way of working helps **reduce errors**, since you can run the script line by line and see what's happening in each step to ensure that things are working properly the way they are meant to.

Allied to the control it provides, knowing a programming language allows you gaining in **efficiency** and **speed**. It may take some time at first to build the required skills to write efficient scripts, but once acquired, it will pay you back exponentially. A simple example showcasing this could be situations in which you have analyzed your data, and either realized that the data should be altered or a different project with similar type of data also needs analyzing. In both scenarios, you would traditionally need to rerun the full set of analyses manually, which can be time-consuming. However, with a programming language, you can update all your tables, figures, and reports by simply applying to the new data your previous scripts.

Such solution brings us to the next reason, which is related to **abstract thinking** and **problem-solving** mindset. These are the two components that are necessary to acquire good programming skills (no worries if you're not confident in having that in you yet; the more you program, the more you'll develop these skills) and thus **increasing your capability** through **continuous improvement**. In other words, the more you play with your data, try new things, etc., the more you'll improve as a programmer, and most importantly, the more diverse and flexible you'll become. And quickly enough, you'll discover that each challenge can be solved in various different ways (as you will see in 4.2.3), so be imaginative and don't be afraid to think outside the box.

Last but not least, it **improves collaboration** and allows for **reproducible research** as your analyses are made transparent to colleagues if you decide to share your scripts with them. By embedding scripts, data sets, and results in a single file (we also recommend adding explanations regarding eventual decisions that were made for clarity), you and your colleagues can always track down why you obtain certain results by simply rereading your script or re-running the analyses. In situations in which multiple users are collaborating on the same project, **version control** (see 2.4) also allows tracking changes done by the different contributors.

### 2.1.3 Why R?

For sensory and consumer scientists, we recommend the R ecosystem for three main reasons.

The first reason is **cultural**. R has from its inception been oriented more toward statistics than to computer science, making the feeling of programming in R more natural (in our experience) for sensory and consumer scientists than Python for instance. This opinion of experience is not to say that a sensory and consumer scientist shouldn't learn other languages (such as Python) if they are inclined to, or even that other tools aren't sometimes better than their R equivalent. Yet, to our experience, R tools are typically better suited to sensory and consumer science than any other solutions we are aware of (especially in programming language).

This leads to our second reason, namely **availability**. R provides many tools that are suitable and relevant for sensory and consumer science purposes, while also providing many packages (e.g. {`SensoMineR`} and {`FactoMineR`}, {`SensR`}, {`FreeSortR`}, {`cata`} just to name a few...) that have been specifically developed for the analysis of sensory and consumer data. If you want to learn more about R, especially in the context of analyzing sensory and consumer data, refer to Lê and Worch (2018).

Finally, the recent work done by the RStudio company, and especially the exceptional work of Hadley Wickham, has lead to a very low barrier to entry for programming within R. This is supplemented by the strong support provided by an active online community via numerous forums and websites, and by the several books, courses, and other educational materials made available.

### 2.1.4 Why RStudio/Posit?

**RStudio** (now renamed as Posit) is a powerful and easy way to interact with R programming. It is an Integrated Development Environment (IDE) for R[1] that comes with a multipanel window setup that provides access to all primary things on a single screen. Such an approach facilitates writing code since all information is available in a single window that includes a console, a script editor that supports direct code execution, as well as tools for plotting, history, debugging, and workplace management (see https://www.rstudio.com/[2]).

Besides the convenience of having all panels on a single screen, we strongly recommend the use of Rstudio as it offers many important features that facilitate scripting. For instance, the script editor provides many features including autocompletion of functions/R elements, hover menus that provide information regarding the arguments of the functions, and handy shortcuts (see Section 2.2.4), etc. Additionally, the Environment section provides easy access to all objects available in the console. Last but not least, RStudio works with a powerful system of *projects* (see 2.2.5).

[1] Originally, RStudio was only developed for R. More recently, it has extended its use for other programming languages (e.g. Python), and to accentuate its reach to other programming languages, RStudio changed its name to Posit to avoid the misinterpretation that it is only dedicated to R.

[2] As we are writing this book, the name Posit is not yet in use, and the website is still defined as *rstudio*.

### 2.1.5 Installing R and RStudio

The first step in this journey is to install R. For this, visit the R Project for Statistical Computing[3] website. From there, follow the download instructions to install R on your operating system. We suggest you download the latest version of R and install it with default options. Note that if you are running R 4.0 or higher, you will need to install Rtools[4].

Next, you need to install RStudio/Posit. To do so, visit the RStudio desktop download page[5] and follow the installation instructions. Download and install the latest version of RStudio with default options.

We then advise you to apply the following adjustments:

- Uncheck *Restore .RData* into the workspace at the startup (*Tools > Global Options... > General*)
- Select *Never* for *Save workspace to .RData on exit* (*Tools > Global Options... > General*)
- Change the color scheme to dark (e.g. "Idle Fingers") (*Tools > Global Options... > Appearance*)
- Put the console on the right (*View > Panes > Console on Right*)

Many other options are available, and we let you explore them yourself to customize Rstudio to your own liking.

## 2.2 Getting Started with R

### 2.2.1 Conventions

Before starting with R, it is important to talk about a few writing conventions that will be used in this book. These conventions are those that are adopted in most books about R.

Throughout this book, since the goal is to teach you to read and write your own code in R, we need to refer to some R functions and R packages. In most cases, the raw R-code that we will be writing and that we advise you to reproduce is introduced in some *special sections* such as:

```
1 + 1
```

```
## [1] 2
```

[3] https://www.r-project.org/
[4] https://cran.r-project.org/bin/windows/Rtools/
[5] https://rstudio.com/products/rstudio/download/

This section shows the code to type on top, and the results (as shown by the R console) in the bottom. To save some space, we may not always show the outputs of the code. Hence it is important for you to run the code to learn it, and to understand it.

Since in most situations, providing code alone is not sufficient, we will also provide explanation in writing. When doing so, we need to refer to R functions and packages throughout the text. In that case, we will clearly make the distinctions between R objects, R functions, and R packages by applying the following rules:

- An R object will be written simply as such: `name_object`
- An R function will always be written by ending with *()*: `name_function()`
- An R package will always be written between {}: {`name_package`}

In some cases, we may want to specify from which package a function belongs to. Rather than calling `name_function()` from the {`name_package`} package, we adopt the R terminology `name_package::name_function()`. This terminology is very important to know and (sometimes) to use in your script to avoid surprises and error.

For illustration, multiple packages have a function called `select()`. Since we are often interested in using the `select()` function from the {`dplyr`} package, we can use `dplyr::select()` in our code to call it. The reason for this particular writing is to avoid errors by calling the *wrong* `select()` function. By simply calling `select()`, we call the `select()` function from the last package loaded that contains a function with that name. However, by specifying the package it belongs to (here {`dplyr`}), we ensure that the right `select()` function (here from {`dplyr`}) is always called.

### 2.2.2 Install and Load Packages

The base installation of R comes with many useful packages that contain many of the functions you will use on a daily basis. However, once you want some more specific analyses, you will quickly feel the urge to extend R's capabilities. This is possible by using R packages.

An R package is a collection of functions, data sets, help files, and documentation, developed by the community that extends the capabilities of base R by improving existing base R functions or by adding new ones.

As of early 2022, there were more than 16,000 different packages available on the CRAN alone (excluding packages that are available through other sources such as GitHub). Here is a short list of packages that we will be consistently using throughout this book.

- Essential packages (or collections): {`tidyverse`}, {`readxl`}, and {`writexl`}.
- Custom Microsoft office document creation: {`officer`}, {`flextable`}, {`rvg`}, and {`openxlsx`}.
- Sensory specific packages: {`SensoMineR`}, {`FactoMineR`}, and {`factoextra`}.

There are many more packages available for statistical tests of all varieties, to multivariate analysis, to machine learning, to text analysis, etc., some being mentioned later in this book.

---

Due to this extensive number of packages, it is not always easy to remember which package does what nor what are the functions that they propose. Of course, the help file can provide such information. More interestingly, some packages provide *Cheat Sheets* that aim to describe the most relevant functions and their uses. Within RStudio, some Cheat Sheets can be found under *Help > Cheat Sheets*, but many more can be found online.

---

To install a package, you can type `install.packages("package_name")` in your console. R will download (an internet connection is required) the packages from the CRAN and install it your computer. Each package only needs to be installed once per R version.

```
install.packages("tidyverse")
```

---

If a script loads a package that is not yet installed, RStudio will prompt a message on top so that you can install them directly. Also, note that if you do not have write access on your computer, you might need IT help to install your packages.

---

Once you have installed a package onto your computer, its content is only available for use once it's loaded. To load a package, use `library (package_name)`.

```
library(tidyverse)
```

A package should only be installed once; however, it should be loaded for each new session of R. To simplify your scripting, we recommend to start your scripts with all the packages that you would need. So, as soon as you open your script, you can run the first lines of code and ensure that all your functions are made available to you.

If you forget to load a package of interest, and yet run your code, you will get an error of the sort: `Error in ...: could not find function "..."`

Note that certain packages may no longer be maintained, and the procedure presented above hence no longer works for those packages. This is, for instance, the case for {`sensR`}, an excellent package dedicated to the analysis of discrimination tests.

```
install.packages("sensR")
```

As you can see, running this code provides the following message: `Warning in install.packages : package 'sensR' is not available for this version of R.`

No worries, there is an alternative way to get it installed by using the `install_version()` function from {`remotes`}. In this case, we need to provide the version of the package to install. Since the latest version of {`sensR`} is 1.5.2, we can install it as following:

```
remotes::install_version("sensR", version = "1.5.2")
```

Last but not least, packages are often improved over time (e.g. through bug fixes, addition of new functions, etc.). To update some existing packages, you can use the function `update.packages()` or simply reinstall it using `install.packages(package_name)`.

---

RStudio also proposes a section called *Packages* (bottom right of your screen if you applied the changes proposed in 2.1.4) where you can see which packages are installed, install new packages, or update already existing packages in a few clicks.

---

### 2.2.3 First Analysis in R

Like any language, R is best learned through examples. Let's start with a simple example where we analyze a tetrad test to illustrate the basic principles.

Suppose you have 15 out of 44 correct answers in a tetrad test. Using the package {sensR},[6] it's very easy to analyze these data:

```
library(sensR)

num_correct <- 15
num_total <- 44

discrim_res <- discrim(correct = num_correct, total = num_total,
                       method = "tetrad")
print(discrim_res)
```

```
##
## Estimates for the tetrad discrimination protocol with 15 correct
## answers in 44 trials. One-sided p-value and 95 % two-sided confidence
## intervals are based on the 'exact' binomial test.
##
##         Estimate Std. Error Lower Upper
## pc        0.3409     0.0715 0.333 0.499
## pd        0.0114     0.1072 0.000 0.249
## d-prime   0.2036     0.9659 0.000 1.019
##
## Result of difference test:
## 'exact' binomial test:  p-value = 0.5141
## Alternative hypothesis: d-prime is greater than 0
```

In a few lines of code, you have just analysed your tetrad test data.

### 2.2.4 R Scripts

You may have entered the code to analyze your tetrad test data directly into the R Console. Although this is possible, and there are many situations where it makes sense (e.g. opening a help menu, taking a quick look at your data, debugging a function, or maybe a simple calculation or testing), it is not the most efficient way of working and we would recommend **NOT** to do so. Indeed, the code directly written in the console cannot be easily modified, retrieved, or saved. Hence, once you close or restart your R session, you will lose it all. Also, if you make an error in your code (even just a typo) or simply want to make a small change, you will have to reenter the entire set of commands, typing it all over again. For all those reasons (and many more), you should write any important code into a **script**.

An R script is simply a text file (with the extension *.R*) containing R code, set of commands (that you would enter on the command line in R) and comments that can easily be edited, executed, and saved later for (re)use.

[6] In the previous section, we showed you how to install it!

You can create a new script in RStudio by clicking the *New File* icon in the upper left corner of the main toolbar and then selecting *RScript*, by clicking *File* in the main menu and then selecting *New File > R Script*, or by simply using *CTRL + SHIFT + N* (Windows)[7]. The script will open in the Script Editor panel and is ready for text entry. Once you are done, you can save your script by clicking the *Save* icon at the top of the Script Editor and open it later to rerun your code and/or continue your work from where you left it.

Unlike typing code in the console, writing code in an R script is not being executed. Instead, you need to send/run it to the console. There are a few ways to do this. If you want to run a line of code, place the cursor anywhere on the line of the code and use the shortcut *CTRL + Enter*. If you want a portion of the code, select by highlighting the code of interest and run it using the same shortcut. To run the entire script (all lines of the code), you can click "Run" in the upper right corner of the main toolbar or use the shortcut *CTRL + SHIFT + Enter*.

A few other relevant shortcuts are:

- Interrupt current command: *Esc*
- Navigate command history: *up and lower arrows*
- Attempt completion: *Tab*
- Call help for a function: *F1*
- Restart R Session: *CTRL + SHIFT + F10*
- Search in File: *CTRL + F*
- Search in All Files (within a project or folder): *CTRL + SHIFT + F*
- Commenting a line of code: *CTRL + SHIFT + C*
- Insertion of a section in the code: *CTRL + SHIFT + R*
- Insertion of a pipe (`%>%`): *CTRL + SHIFT + M*

There are many more shortcut options. A complete list is available within R Studio under *Tools > Keyboard Shortcut Help* (or directly using *ALT + SHIFT + K*). So have a look at them, and don't hesitate to learn by heart the one that you use regularly as it will simplify your scripting procedure.

### 2.2.5 Create a Local Project

Next to scripts, working with RStudio projects will facilitate your life even further. RStudio projects make it straightforward to divide your work into multiple contexts, each with its own working directory, workspace, history, and source documents. It keeps all of your files (R scripts, R markdown documents,

[7] The shortcuts are given for Windows users. For Mac users, replace *CTRL* by *Cmd* and it should also work.

R functions, data, etc.) in one place. RStudio projects allow independence between projects, which means that you can open more than one project at the time and switch at ease between them without fear of interference (they all use their own R session). Moreover, those projects are not linked to any computer, meaning that the file path are linked to the project itself: While sharing a RStudio project with colleagues, they do not need to update any file path to make it work.

To create a new project locally in RStudio, select *File* > *New Project* from the main menu. Typically, a new project is created in a new directory, and you can also transform an already existing folder on your computer into an RStudio Project. You can also create a new project by clicking on the *Project* button in the top right of RStudio and selecting *New Project...*. Once your new project has been created, you will now have a new folder on your computer that contains the basic file structure. You probably want to add folders to better organize all the files and documents, such as folders for input, output, and scripts.

For consistency, we suggest you keep the same folder structure across projects. For example, you may create a folder that contains your scripts, one for the data and one for exporting results from R (excel files, figures, report, etc.). If you adopt this strategy, you may see an interest in the code below, which automatically creates all your folders. To run this code, the {`fs`} package is required. Here, 5 folders are being created:

```
library(fs)

fs::dir_create(path=c("code", "data", "docs", "output", "template"))
```

## 2.3 Further Tips on How to Read This Book?

In this book, we assume that the readers have already some basic knowledge in R. If you are completely new to R, we recommend you reading "R for Data Science" by Wickham and Grolemund[8] (2016) or looking at some documentation online to get you started with the basics.

Just like with any spoken language, the same message can be said in various ways. The same applies with writing scripts in R, each of us having our own styles or our own preferences toward certain procedures, packages, functions, etc. In other words, writing scripts is personal. Through this book, we are not trying to impose our way of thinking/proceeding/building scripts, instead we

[8] https://r4ds.had.co.nz/

aim at sharing our knowledge built through past experiences to help you find your own.

But to fully decode our message, you'll need some reading keys. These keys will be described in the next sections.

Note that the lines of code presented in this section do not run and are simply presented for illustration.

### 2.3.1 Introduction to {magrittr} and the Notion of Pipes

R is an *evolving* programming language that expands very rapidly.

If most additions/improvements have a fairly limited reach, the introduction of the {tidyverse} in 2016 by H. Wickham revolutionized the way of scripting in R for many users. At least for us, it had a large impact as we fully embraced its philosophy, as we see its advantage for Data Science and for analyzing our sensory and consumer data. It is, hence, no surprise that you'll read and learn a lot about it in this book.

As you may know, {tidyverse} is a grouping of packages dedicated to Data Science, which includes (among others) {readr} for data importation, {tibble} for data structure, {stringr} and {forcats} for handling strings and factors, {dplyr} and {tidyr} for manipulating and tidying data, {ggplot2} for data visualization, and {purrr} for functional programming. But more importantly, it also includes {magrittr}, the package that arguably impacted most of our way of scripting by introducing the notion of *pipes* (defined as `%>%`) as it provides code that is much easier to read and understand.

To illustrate the advantage of coding with pipes, let's use the example provided by H. Wickham in his book *R for Data Science*. It is some code that tells a story about a little bunny names Foo Foo: > Little bunny Foo Foo > Went hopping through the forest > Scooping up the field mice > and bopping them on the head.

If we were meant to tell this story though code, we would start by creating an object name `FooFoo` which is a little bunny:

```
foo_foo <- little_bunny()
```

To this object, we then apply different functions (we save each step as a different object):

```
foo_foo_1 <- hop(foo_foo, through=forest)
foo_foo_2 <- scoop(foo_foo_1, up=field_mice)
foo_foo_3 <- bop(foo_foo_2, on=head)
```

One of the main downsides of this approach is that you'll need to create intermediate names for each step. If natural names can be used, this will not be a problem, otherwise it can quickly become a source of error (using the wrong object for instance)! Additionally, such an approach may affect your disk memory since you're creating a new object in each step. This can be problematic when the original data set is large.

As an alternative, we could consider running the same code by over-writing the original object:

```
foo_foo <- hop(foo_foo, through=forest)
foo_foo <- scoop(foo_foo, up=field_mice)
foo_foo <- bop(foo_foo, on=head)
```

If this solution looks neater and more efficient (less thinking, less typing, and less memory use), it is more difficult to debug, as the entire code should be rerun from the beginning (when `foo_foo` was originally created). Moreover, calling the same object in each step obscures the changes performed in each line.

To these two approaches, we prefer a third one that strings all the functions together without intermediate steps of saving the results. This procedure uses the so-called pipes (defined by `%>%`), which takes automatically as input the output generated by the previous line of code:

```
foo_foo %>%
  hop(through = forest) %>%
  scoop(up = field_mice) %>%
  bop(on = head)
```

This code is easier to read and understand as it focuses more on the verbs (here `hop()`, `scoop()`, and `bop()`) rather than the names (`foo_foo_1` or `foo_foo`). It can be surprising at first, but no worries, by the time you've read this book, you'll be fully familiar with this concept.

When lines are piped, R runs the entire block at once. So how can we understand the intermediate steps that were done or how can we fix the code if an error occurs? The answer to these questions is simple: run back the code bits by bits.

For instance, in this previous example, we could start by printing `foo_foo` (in practice, only select `foo_foo` and run this code only) only to ensure that it is the object that we were supposed to have. If it is the case, we can then extend the selection to the next line by selecting all the code until (but excluding[9]!) the pipe. Repeat this until you find your error or you've ensured that all the steps have been performed correctly.

While reading this book, we advise you to apply this trick to each long pipe for you to get a hand on it and to visualize the intermediate steps.

---

Within a pipe, it is sometime needed to call the temporary data or output generated in the previous step. Since the current object does not exist yet nor have a name (it is still *under-construction*), we need to find another way to call it. In practice, this is very simple and can be done by using . as we will see it extensively in Chapter 4.

---

Note however that although pipes are very powerful, they are not always the best option:

- A rule of thumb suggests that if you are piping more than 10 lines of code, you're probably better of splitting it into 2 or more blocks (saving results in intermediate step) as this simplifies debugging.
- If some steps require multiple inputs or provides multiple outputs, pipes should not be used as they usually require a primary object to transform.
- The system of pipes works linearly: if your code requires a complex dependency structure, then the pipes should be avoided.

### 2.3.2 Tibbles

Within the {`tidyverse`}, another valuable (yet often forgotten) package is the {`tibble`} package. This package aims at providing a new format to store data.

In appearance, a *tibble* looks like just any other table, whether it is a matrix or a data frame. But in the background, it is defined as an optimized version of a data frame that kept (and extended) all the relevant parts and removed all the unnecessary parts.

To show you the properties of a tibble, let's load the data set called `sensochoc` from the `SensoMineR` package:

```
library(tidyverse)
library(SensoMineR)
data(chocolates)
```

[9] If your code ends up on a pipe, R is expecting additional code and will not show results. This usually creates errors since the next piece of code is probably not matching the current pipe's expected code.

If you type sensochoc in your R session, the entire data set will be printed, which makes it difficult to read. Here, we opt for a simpler solution by only showing the first lines using head()

---

Here and throughout the book, some of the outputs are being reduced and only the first results are being printed (to avoid printing too much things that may not be so relevant.). So do not panic if only a part of the outputs is shown in the book compared to your screen.

---

```
head(sensochoc)
```

```
##      Panelist Session Rank Product CocoaA MilkA
## I001        1       1    1   choc6      7     6
## I002        1       1    6   choc3      6     7
## I003        1       1    3   choc2      8     6
## I004        1       1    5   choc1      7     8
## I005        1       1    2   choc4      8     5
## I006        1       1    4   choc5      7     5
```

This is the typical output from a matrix() or data.frame() in R. In particular, one can note that the first column does not have a header as it represents the row names.

Let's convert this table into a tibble (using as_tibble()) and look at the output:

```
sensochoc_tb <- sensochoc %>%
  as_tibble()
print(sensochoc_tb)
```

```
## # A tibble: 348 x 18
##   Panelist Session Rank  Product CocoaA MilkA CocoaF
##   <fct>    <fct>   <fct> <fct>    <int> <int>  <int>
## 1 1        1       1     choc6        7     6      6
## 2 1        1       6     choc3        6     7      2
## 3 1        1       3     choc2        8     6      5
## 4 1        1       5     choc1        7     8      8
## # ... with 344 more rows, and 11 more variables:
## #   MilkF <int>, Caramel <int>, Vanilla <int>,
## #   Sweetness <int>, Acidity <int>, Bitterness <int>,
## #   Astringency <int>, Crunchy <int>, Melting <int>,
## #   Sticky <int>, Granular <int>
```

The appearance of the table looks quite different, and there are some interesting things that can be noticed:

- The dimensions of the table are shown straight at the top.
- By default, the printing function for tibbles only prints a predefined number of rows (by default 10) and as many columns as the screen allows. All other columns are informed in some text under the table.
- Although data frame converts some names (e.g. replacing spaces with .), tibbles keep the same as in the original file.
- Numbers are also automatically formatted to 3 values after the decimal (by default), and negative values are printed in red.
- Under the header, the type of each variable is informed. Although very valuable, this information is not provided with matrix or data frame.
- The row names are lost, as tibbles do not have row names. In this example, there is no information in the row names, so we can ignore them, although we could have easily recovered them by adding `rownames = "name variable"` in `as_tibble()` which would have added them as a new column called `name variable` to the data.

---

Certain packages including {`SensoMineR`} do not accept tibbles. Instead, they require matrices or data frames as inputs, and use their row names for certain analyses (see Section 10.1 for an example of PCA). Fortunately, there is a very easy way to convert tibbles to (say) data frame through `as.data.frame()` combined with `column_to_rownames()` to automatically pass the information present in a column as row names (the complementary function `rownames_to_column()` also exists).

---

To go further, let's extract from `sensochoc` the 4th column (`Product`):

```
sensochoc[,4]
```

```
## [1] choc6 choc3 choc2 choc1 choc4
## Levels: choc1 choc2 choc3 choc4 choc5 choc6
```

By extracting a column of a data frame, the resulting output is converted into a vector. Let's reproduce the same extraction to `sensochoctb`:

```
sensochoc_tb[,4]
```

```
## # A tibble: 348 x 1
##   Product
##   <fct>
## 1 choc6
## 2 choc3
## 3 choc2
## 4 choc1
## # ... with 344 more rows
```

Subsetting from a tibble with [] always returns a tibble, which is very convenient with programming as we then know what to expect (unlike data frame which can return a data frame or a vector depending on the situation).

Last but not least, tibbles can take as entries single elements (e.g. numbers, characters, dates, etc.), but also lists of elements. This particular property is very interesting since it allows combining different outputs in one table although they may have different structures.

Let's start with converting the table in a long format:

```
sensochoc_tb <- sensochoc_tb %>%
  dplyr::select(-c("Session","Rank")) %>%
  pivot_longer(-c("Panelist","Product"), names_to="Attributes",
               values_to="Scores")
```

This new tibble has 4872 rows and 4 columns. Since we have 14 attributes, that means that each attribute has 348 data points.

Let's nest the results by `Attributes`:

```
sensochoc_tb <- sensochoc_tb %>%
  nest_by(Attributes)
```

This new tibble has only 14 rows (one per attribute), yet the new column called `data` contains a list of dimensions [348 rows and 3 columns]. This corresponds to the original data.[10]

Let's now run an ANOVA for each attribute:

```
sensochoc_tb <- sensochoc_tb %>%
  mutate(ANOVA = list(aov(Scores ~ Product + Panelist, data=data)))
```

[10] We could easily retrieve the original data by simply using `sensochoc_tbl %>% unnest(data)`.

This code adds another column (called ANOVA) that contains the results of the ANOVA model. Let's imagine we're interested in Acidity; Then, we could extract the results of the anova as follows:

```
sensochoc_tb %>%
  filter(Attributes == "Acidity") %>%
  summarize(broom::tidy(ANOVA))
```

```
## Warning: Returning more (or less) than 1 row per 'summarise()' group was deprecated in dplyr 1.1.0.
## i Please use 'reframe()' instead.
## i When switching from 'summarise()' to 'reframe()', remember that 'reframe()' always returns an
##   ungrouped data frame and adjust accordingly.
```

```
## 'summarise()' has grouped output by 'Attributes'. You
## can override using the '.groups' argument.
```

```
## # A tibble: 3 x 7
## # Groups:   Attributes [1]
##   Attributes term     df sumsq meansq stati~1   p.value
##   <chr>      <chr> <dbl> <dbl>  <dbl>   <dbl>     <dbl>
## 1 Acidity    Prod~     5  325.  65.0    17.1   5.67e-15
## 2 Acidity    Pane~    28  917.  32.7     8.62  2.88e-25
## 3 Acidity    Resi~   314 1193.   3.80   NA    NA
## # ... with abbreviated variable name 1: statistic
```

Since extracting the information regarding the models is also of interest, let's add that to our tibble as well. This can easily be done using the glance() function from {broom}:

```
sensochoc_tb <- sensochoc_tb %>%
  mutate(Results = list(broom::glance(ANOVA)))
```

Here again, if we would want to see the results for Acidity only, then we could extract that information as follows:

```
sensochoc_tb %>%
  filter(Attributes == "Acidity") %>%
  summarize(Results)
```

```
## 'summarise()' has grouped output by 'Attributes'. You
## can override using the '.groups' argument.
```

```
## # A tibble: 1 x 7
## # Groups:   Attributes [1]
##   Attributes logLik   AIC   BIC deviance  nobs r.squa~1
##   <chr>       <dbl> <dbl> <dbl>    <dbl> <int>    <dbl>
## 1 Acidity     -708. 1486. 1621.    1193.   348    0.510
## # ... with abbreviated variable name 1: r.squared
```

So as we can see, the same tibble of 14 row and 4 columns contain, per attribute, the raw `data`, the results of the `ANOVA`, as well as the overall `Results` of each of the model. Although all these tables have completely different structures (`data` has 348 rows and 3 columns, whereas `Results` has 1 row and 6 columns), they are still related to the same *objects* (here attributes). Hence, keeping them in the same *tidy* place facilitates follow-up analysis by avoiding creating multiple objects and reducing the risks of error. An example of such use of tibbles is provided in Section 10.3, in which we also show how to use the information stored in different elements.

### 2.3.3 Calling Variables

In R, variables can be called in different ways when programming. If the names of variables should be read from the data (e.g. "Product", "products", "samples", etc.), you will often use strings, meaning that the name used will be defined between quotes (e.g. `"Product"`).

Within the {`tidyverse`}, the names of variables that are included within a data set are usually called as it is, without quote:

```
sensory %>%
  dplyr::select(Judge, Product, Shiny)
```

This is true for simple names that do not contain any special characters (e.g. space, `-`, `etc.`). For names that contain special characters, the use of *backticks* are required (note that *backticks* can also be used with simple names):

```
sensory %>%
  dplyr::select(`Judge`, Product, `Color evenness`).
```

While going through this book, you'll notice that many functions from the {`tidyverse`} sometimes require quotes, and sometimes don't. The simple way to know whether quotes are required or not is based on its existence in the data set or not: If the column exists and should be used, no quotes should be used. On the contrary, if the variable doesn't exist and should be created, then quotes should be used.

Let's illustrate this through a simple example involving `pivot_longer()` and `pivot_wider()` successively (see 4.2.2 for more information). For `pivot_longer()`, we create two new variables, one that contains the column names (informed by `names_to`) and one that contains the values (informed by `values_to`). Since these variables are being created, quotes are required for the new names. For `pivot_wider()`, quotes are not needed since the names of the variables to use (`names_from` and `values_from`) are present in the data:

```
sensory %>%
  pivot_longer(Shiny:Melting, names_to="Variables",
               values_to="Scores") %>%
  pivot_wider(names_from=Variables, values_from=Scores)
```

Unfortunately, this rule of thumb is not always true (e.g. `separate()`, `unite()`, and `column_to_rownames()`) but you'll quickly get familiar with these exceptions.

### 2.3.4 Printing vs. Saving Results

In many examples through this book, we apply changes to certain elements without actually saving them in an R object. This is quite convenient for us as many changes we do are only done for pedagogic reasons, and are not necessarily relevant for our analyses.

Here is an example of such a case (see Section 4.2.1):

```
sensory %>%
  rename(Panellist = Judge, Sample = Product)
```

When you run this code, you can notice that we rename `Judge` to `Panellist`, and `Product` to `Sample`... at least this is what you see on screen. However, if you look at `sensory`, the data set still contains the column `Judge` and `Product` (`Panellist` and `Sample` do not exist!). This is simply because we did not save the changes.

If we would want to save the element in a new object, we should save the outcome in an element using <-:

```
newsensory <- sensory %>%
  rename(Panellist = Judge, Sample = Product)
```

Here, `newsensory` corresponds to `sensory`, but with the new names. Of course, if you would want to overwrite the previous file with the new names, you simply need to ensure that the name of the output is the same as the name of the input (like we did with `foo_foo` in 2.3.1). Concretely, we replace here `newsensory` by `sensory`, meaning that the new names are saved in `sensory` (so the old names `Judge` and `Product` are definitely lost). This procedure saves computer memory and does not require you coming up with new names all the time. However, it also means that some changes that you applied may be lost, and if you have a mistake in your code, it is more complicated to find and ultimately solve it (you may need to rerun your entire script).

```
sensory <- sensory %>%
  rename(Panellist = Judge, Sample = Product)
```

To visualize the changes, you would need to type `newsensory` or `sensory` in R. Another (faster) way to visualize it is to put the entire block of code between brackets: Putting code between brackets is equivalent to asking to print the output after being run.

```
(sensory <- sensory %>%
  rename(Panellist = Judge, Sample = Product))
```

Note that if you run all these lines of codes in R, you will get an error stating `Column 'Judge' doesn't exist`. This is a good illustration of a potential error mentioned above: We overwrote the original `sensory` (containing `Judge` and `Product`) with another version in which these columns were already renamed as `Panellist` and `Sample`. So when you rerun this code, you are trying to apply again the same changes to columns that no longer exist, hence the error.

This is something that you need to take into consideration when overwriting elements (in this case, you should initialize `sensory` to its original version before trying).

---

It is also worth reminding the readers that we deliberately **do not print** each and every output generated. Instead, we advise the reader to rewrite the code and run it in parallel as they read the book. The outputs will then be printed on their screen. Also, in some cases, we took the liberty to reduce the outputs (e.g. only showing the first rows and/or columns) which can lead to some inconsistencies between what is printed in the book and what is shown on your screen. Please be aware that this can (and will) happen.

---

### 2.3.5 Running Code and Handling Errors

For you to get the most out of this book, you need to understand (and eventually adhere to) our philosophy of scripting and our way of working. This is why we are providing you with some tips to use, if you're comfortable with them:

1. Create a folder for this book on your computer and create a script for each chapter in which you retype yourself each line of code. If you work with the online version, you could copy/paste the code to go faster, but you may miss some subtleties.
2. Do not be discouraged when you get some errors: we all get some. At first, this can be very frustrating, especially when you are not able to fix them quickly. If you get stuck on an error and cannot fix it immediately, take a break and come back later with fresh eyes, and you may solve it then. And with time and experience, you'll notice that you can reduce the amount of errors and will also solve them faster (you will also learn to understand the error messages provided by R).
3. The more code, the more difficult it is to find errors. This is true whether you use *regular* R-code or pipes. The best way to solve errors in such circumstances is to run the code line by line until you find the error and understand why the input/output does not match the expectations.
4. In the particular case of pipes, debugging errors means that you shouldn't run the entire block of code, but select parts of it and run it by adding in each run a new line. This can either be done by stopping your selection just before the adequate `%>%` sign (as mentioned earlier) or by adding after the last `%>%` sign the function `identity()`.[11]

[11] `identity()` is a function that returns as output the input as it is. This function is particularly useful in pipes as you can finish your pipes with it, meaning that you can put any line in comments (starting with '#') without worrying about finishing your pipe with a `%>%`.

## 2.4 Version Control/Git and GitHub

Version control is a tool that tracks changes to files, especially source code files. Using version control means that you can not only track the changes but manage them by, for instance, describing the changes or reverting to previous versions. This is particularly important when collaborating with other developers.

Version control systems are simply software that helps users manage changes to source code over time. The reasons why everyone should use version control include backing up work, restoring prior versions, documenting reasons for changes, quickly determining differences in versions, easily sharing code, and developing in parallel with others.

There are many tools for Version Control out there, but Git/GitHub are by far the most common ones. We highly recommend that you integrate both Git and GitHub into your data science workflow. For a full review of Git and GitHub from an R programming perspective, we recommend Happy Git with R[12] by Jenny Bryant. In what follows, we simply provide the minimum information needed to get you up and running with Git and GitHub. Also, for an insightful discussion of the need for version control, please see Bryan (2018).

### 2.4.1 Git

Git is a version control system that runs locally and automatically organizes and saves versions of code on your computer, but does not connect to the internet. Git allows you to revert to earlier versions of your code, if necessary. To set up Git, follow the following steps:

1) Download and install the latest version of Git. Download and install Git with standard options (allow third party software) for Windows[13] or Mac[14]

2) Enable Git Bash in RStudio: Go to "Tool" on the top toolbar and select "Global Options..." > "Terminal". In the drop-down box for "New terminals open", select "Git Bash".

3) Configure Git from Rstudio: The easiest way is to use the package `{usethis}`

[12] https://happygitwithr.com/
[13] https://git-scm.com/download/win
[14] https://git-scm.com/download/mac

```
library(usethis)
use_git_conf (user.name = "your username",
              user.email = "your email address")
```

### 2.4.2 GitHub

GitHub is a cloud-based service that supports Git usage. It allows online backups of your code and facilitates collaboration between team members. While Git creates local repositories on your computer, GitHub allows users to create remote online repositories for their code.

To set up GitHub, follow the steps below:

1) Register for a GitHub Account: To get started you can sign up for a free GitHub account: GitHub[15]

We recommend you not to tie your account to your work email and to use all lowercase to avoid confusion.

2) Create a Test Repository in GitHub

Once you log into your account, create a new repository by clicking the green button "New". You have to then name your repository and make some selections. We recommend you select the option "Private" and click on the option "Initialize this repository with a README". The last step is to click on "Create Repository".

Once the repository has been created you need to copy the repository URL to create a project in RStudio (next step). If you select the repository you just created, click on the green button "Code" and copy the URL link.

3) Create an RStudio Project from GitHub

As we have seen, to create a new project, select "File" > "New Project..." from the top bar menu or by clicking on the "Project" button in the top right of RStudio and by selecting "New Project...". Select then "Version Control" > "Git". Paste the repository URL link, select where you want to save this project locally, and click "Open in new session". Finally, click "Create Project".

4) Register GitHub from Studio

[15] https://github.com/

At this point, you will be asked to log into GitHub from RStudio. You should only have to do this once.

5) Push and Commit Changes

Once you are done with your coding, or have finished updating a series of scripts, you can simply push or send them to GitHub, so others can see your changes. You have to first commit and then push it to GitHub. To do so, you can click the "Git" icon on the top menu of RStudio and select the option "Commit". You can select what you want to commit and describe the changes you did. After committing your code/files, you have to push it by clicking the option "Push".

6) Pull Changes

In case you are working with other colleagues, a good practice is to always pull (which means download) the latest code available (i.e. the code that your collaborators have recently pushed) before you get started and before pushing any changes. To do so, you can click the "Git" icon on the top menu and select the option "Pull".

If you've read this through (no worries if everything is not completely clear yet, it will come!), and followed the different steps here, you should be ready to learn data science for sensory and consumer scientists. Let's get started?

# 3

# *Why Data Science?*

In this chapter, we explain what is data science and discuss why data science is valuable to sensory and consumer scientists. While this book focuses on the aspects of data science that are most important to sensory and consumer scientists, we recommend the excellent book from Wickham and Grolemund (2016) for a more general introduction to data science.

## 3.1 History and Definition

You may have heard that data science was called the "sexiest job of the 21st century" by Harvard Business Review (Davenport and Patil, 2012). But what is data science? Before we give our definition, we provide some brief history for context. For a comprehensive survey of this topic, we recommend Cao (2017).

To begin, there was a movement in early computer science to call their field "data science." Chief among the advocates for this viewpoint was Peter Naur, winner of the 2005 Turing award.[1] This viewpoint is detailed in the preface to his 1974 book, "Concise Survey of Computer Methods," where he states that data science is "the science of dealing with data, once they have been established" (Naur, 1974). According to Naur, this is the purpose of computer science. This viewpoint is echoed in the statement, often attributed to Edsger Dijkstr, that "Computer science is no more about computers than astronomy is about telescopes."

Interestingly, a similar viewpoint arose in statistics, as reflected in John Tukey's statements that "Data analysis, and the parts of statistics which adhere to it, must ... take on the characteristics of science rather than those of mathematics" and that "data analysis is intrinsically an empirical science" (Tukey, 1962). This movement culminated in 1997 when Jeff Wu proposed during his inaugural lecture upon becoming the chair of the University of Michigan's statistics department, entitled "Statistics = Data Science?," that statistics should be called data science (Wu, 1997).

[1] A prize roughly equivalent in prestige to a Nobel prize, but for computer science.

DOI: 10.1201/9781003028611-3

These two movements[2] came together in 2001 in William S. Cleveland's paper "Data Science: An Action Plan for Expanding the Technical Areas in the Field of Statistics" (Cleveland, 2001). In this highly influential monograph, Cleveland makes the key assertion that "The value of technical work is judged by the extent to which it benefits the data analyst, either directly or indirectly."

A more recent development in the history of data science has been the realization that the standard outputs of data science such as tables, charts, reports, dashboards, and even statistical models – can be viewed as tools that must be used in the real world in order to be valuable. This realization stems from the influence of the technology sector, where the field of design has focused on improving the ease of use of websites, apps, and devices. To quote Steve Jobs, perhaps the most influential champion of design within the technology space: "Design is not just what it looks and feels like. Design is how it works."

Based on this history, we provide our definition of **data science**:

---

Data science is the intersection of statistics, computer science, and industrial design.

---

Accordingly, we use the following three definitions of these fields:

- **Statistics**: The branch of mathematics dealing with the collection, analysis, interpretation, and presentation of masses of numerical data.
- **Computer Science**: Computer science is the study of processes that interact with data and that can be represented as data in the form of programs.
- **Industrial Design**: The professional service of creating and developing concepts and specifications that optimize the function, value, and appearance of products and systems for the mutual benefit of both user and manufacturer.

Hence data science is the delivery of value through the collection, processing, analysis, and interpretation of data.

[2] It is worth noting that these two movements were connected by substantial work in the areas of statistical computing, knowledge discovery, and data mining, with important work contributed by Gregory Piatetsky-Shapiro, Usama Fayyad, and Padhraic Smyth among many others. See Fayyad et al. (1996), for example.

## 3.2 Benefits of Data Science

Now that we have a working definition of data science, we consider some reasons for sensory and consumer scientists to embrace it. Many of these reasons apply to any modern scientific discipline, yet the fact that sensory and consumer scientists often occupy a central location in their organizations (such as sitting between product development and marketing, for example) means that sensory and consumer scientists must routinely create useful outputs for consumption by a wide variety of stakeholders. Moreover, sensory and consumer data are often diverse, so facility in data manipulation and flexibility in data analysis are especially important skills for sensory scientists.

### 3.2.1 Reproducible Research

One of the most important ideas in data science is that of reproducible research (cf. Peng, 2011). Importantly, reproducibility in the context of data science does not refer to the repeatability of the experimental results themselves if the experiment were to be conducted again. What is instead meant by reproducible research is the ability to proceed from the input data to the final results in reproducible steps. Ideally, these steps should be well-documented so that any future researcher, including the researcher who originally conducted the work, should be able to determine all choices made in data cleaning, manipulation, and analysis that led to the final results. Since sensory and consumer scientists often work in teams, this clarity ensures that anyone on the team can understand the steps that led to prior results and apply those steps to their own research going forward.

### 3.2.2 Standardized Reporting

Related to the idea of reproducible research is that of standardized reporting. By following a data-scientific workflow, including automated reporting (see Chapter 6), we can standardize our reporting across multiple projects. This standardization has many benefits:

- **Consistent Formatting** When standardized reporting is used, outputs created by a team are formatted consistently regardless of who creates them. This consistency helps consumers of the reports – whether those consumers are executives, clients, or other team members to quickly interpret results.
- **Upstream Data Consistency** Once a standardized workflow is put in place, consistency of data formatting gains a new importance as producers of the report can save significant time by not having to reformat new

data. This fact puts pressure on the data collection procedure to become more consistent, which ultimately supports knowledge management.

- **Shared Learning** Once a team combines standardized reporting with tools for online collaboration such as GitHub (see Appendix 2.4), any improvement to reporting (e.g., to a table, chart, text output, or even to the reporting format itself) can be leveraged by all members of the team. Thus improvements compound over time, to the benefit of all team members.

## 3.3 Data Scientific Workflow

A schematic of a data scientific workflow is shown in Figure 3.1 (Diagram inspired by Wickham and Grolemund, 2016). Each section is described in greater detail below.

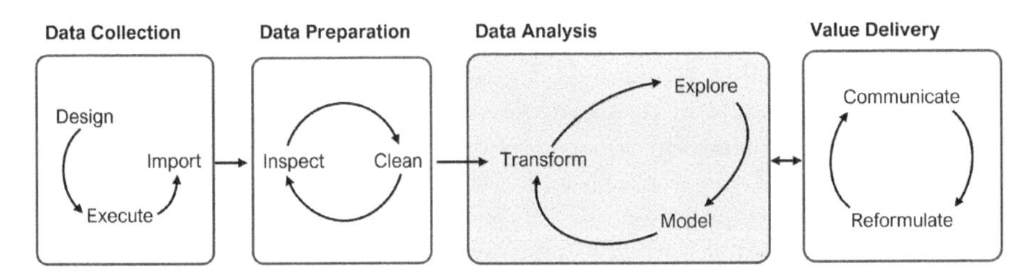


**FIGURE 3.1**
Data scientific workflow.

### 3.3.1 Data Collection

#### *Design*

From the standpoint of classical statistics, experiments are conducted to test specific hypotheses and proper experimental design ensures that the data collected will allow hypotheses of interest to be tested (c.f. Fisher, 1935). Sir Ronald Fisher, the father of modern statistics, felt so strongly on this topic that he said: "To call in the statistician after the experiment is done may be no more than asking him to perform a postmortem examination: he may be able to say what the experiment died of."

This topic of designed experiments, which are necessary to fully explore causal or mechanistic explanations, is covered extensively in Lawson (2014).

Since Fisher's time, ideas around experimental design have relaxed somewhat, with Tukey (1977) arguing that exploratory and confirmatory data analyses can and should proceed in tandem.

Unless exploratory data analysis uncovers indications, usually quantitative ones, there is likely to be nothing for confirmatory data analysis to consider.

Experiments and certain planned inquires provide some exceptions and partial exceptions to this rule. They do this because one line of data analysis was planned as a part of the experiment or inquiry. *Even here, however, restricting one's self to the planned analysis – failing to accompany it with exploration – loses sight of the most interesting results too frequently to be comfortable.* (Emphasis original)

In this book, we take no strong opinions on this topic, as they belong more properly to the study of statistics than to data science. However, we agree that results from an experiment explicitly designed to test a specific hypothesis should be viewed as more trustworthy than results incidentally obtained. Moreover, as we describe in Chapter 12, well-selected sample sets support more generalizable predictions from machine learning models.

#### *Execute*

Execution of the actual experiment is a crucial step in the data science workflow, although not one in which data scientists themselves are necessarily involved. Even so, it is imperative that data scientists communicate directly and frequently with experimenters so that nuances of the data are properly understood for modeling and interpretation.

#### *Import*

Once the data are collected, they need to find their way into a computer's working memory to be analyzed. This importation process should be fully scripted in code, as we detail in Chapter 8, and raw data files should never be directly edited. This discipline ensures that all steps taken to import the data will be understood later and that the reasoning behind all choices will be documented. Moreover, writing code to import raw data allows for new data to be analyzed quickly in the future as long as the data formatting is consistent. For sensory scientists, who regularly run similar tests, a streamlined workflow for data import and analysis both saves much time and protects against errors.

### 3.3.2 Data Preparation

Preparing data for analysis typically involves two steps: data inspection and data cleaning.

#### *Inspect*

In this step, the main goal is to gain familiarity with the data. Under ideal circumstances, this step includes reviewing the study documentation, including the study background, sampling, design, analysis plan, screener (if any), and questionnaire. As part of this step, the data should be inspected to ensure they have been imported properly and relevant data quality checks, such as checks for consistency and validity, should be performed. Preliminary summary tables and charts should also be preformed at this step to help the data scientist gain familiarity with the data. These steps are discussed in further detail in Section 9.2 of Chapter 9.

#### *Clean*

Data cleaning is the process of preparing data for analysis. In this step, we must identify and correct any errors and ensure the data are formatted consistently and appropriately for analysis. As part of this step, we will typically tidy our data, a concept covered in more detail in Section 4.2. It is extremely important that any changes to the data are made in code with the reasons for the changes clearly documented. This way of working ensures that, a year from now, we don't revisit our analysis to find multiple versions of the input data and not know which version was the one used for the final analysis.[3] We discuss data cleaning in further detail in Section 9.3.

### 3.3.3 Data Analysis

Data analysis is one of the areas of data science that most clearly overlaps with traditional statistics. In fact, any traditional or computational statistical technique can be applied within the context of data science.

In practice, the dominant cultural difference between the two fields can be summarized as:

- Statistics often focuses on advancing explicit theoretical understanding of an area through parameter estimation within first-principle models.
- Data science often focuses on predictive ability using computational models that are validated empirically using held-out subsets of the data.

Another cultural difference between the two fields is that data science, evolving more directly out of computer science, has been more historically interested in documenting the code used for analysis with the ultimate goal of reproducible research. See Peng (2011) for more information on this topic,

[3] Anyone working in the field for more than five years has almost certainly experienced this problem, perhaps even with their own data and reports.

for example. This difference is gradually disappearing, however, as statistics more fully embraces a data scientific way of scripting analyses.

Data analysis is covered in greater detail in Chapter 10. The typical steps of data analysis are data transformation, exploration, and modeling, which we review below.

### *Transform*

Data transformation is slightly different from data preparation. In data preparation, we prepare the raw data for processing in a non-creative way, such as reshaping existing data or storing character strings representing dates as date formatted variables. With data transformation, we create new data for analysis by applying functions to the raw data. These functions can be simple transformations, such as inversions or logarithms, or can be summary operations such as computing means and variances, or could be complex operations such as principle components analysis or missing value imputations. In a machine learning context (see Chapter 12), this step is often referred to as "feature engineering." In any case, these functions provide the analyst an opportunity to improve the value of the analysis through skillful choices. Data transformation is covered in more detail in Chapter 10.

### *Explore*

Just as data transformation differs slightly from data preparation, data exploration differs slightly from data inspection. When we inspect the data, our goal is to familiarize ourselves with the data and potentially spot errors as we do so. With data exploration, our goal is to begin to understand the results of the experiment and to allow the data to suggest hypotheses for follow-up analyses or future research. The key steps of data exploration are graphical visualizations (covered in Chapter 5) and exploratory analyses (covered in Chapter 10). As we will discuss later in this book, employing automated tools for analysis requires caution; the ease with which we can conduct a wide range of analyses increases the risk that chance results will be regarded as meaningful. In Chapter 12, we will discuss techniques, such as cross-validation, that can help mitigate this risk.

### *Model*

At last we reach the modeling step of our workflow, which is the step in which we conduct formal statistical modeling. This step may also include predictive modeling, which we cover in Chapter 12, as mentioned above. One difference between data science and classical statistics is that this step may feed back into the transform and explore steps, as data scientists are typically more willing to allow the data to suggest new hypotheses for testing (recall Tukey's quotation above). This step is described in further detail in Chapter 10.

### 3.3.4 Value Delivery

We now arrive at the final stage of the data science workflow, value delivery, which is the stage most influenced by industrial design. Recall the definition we provided above:

- **Industrial Design**: The professional service of creating and developing concepts and specifications that optimize the function, value, and appearance of products and systems for the mutual benefit of both users and manufacturers.

From this perspective, our product consists of the final results as provided to the intended audience. Consequently, we may need to adjust both the results themselves and the way they are presented according to whether the audience consists of product developers, marketing partners, upper management, or even the general public. Hence, in this stage, we communicate our results and potentially reformulate our outputs so that they will provide maximum value to the intended audience. Although we describe value delivery in more detail in Chapter 11, we briefly review the two steps of value delivery, communicate and reformulate, below.

#### *Communicate*

The goal of the communication step is to exchange information stemming from our data scientific work. Importantly, communication is a two-way street, so it is just as important to listen in this step as it is to share results. Without feedback from our audience, we won't be able to maximize the impact of our work. We discuss this topic in more detail in Section 11, and note that automated reporting, which we cover in Chapter 6 also plays a large role in this step by saving time in building slides that can later be spent in thinking about the storytelling aspects of our communications.

#### *Reformulate*

In the final step of our data scientific workflow, we incorporate feedback received during the communication step back into the workflow. This step may involve investigating new questions and revising the way we present results. Since we seek to work in a reproducible manner, the improvements we make to our communication can be committed to code and the lessons these improvements reflect can be leveraged again in the future. It is also important to note that, as we reformulate, we may need to return all the way to the data cleaning step, if we learn during the communication step that some aspects of data import or initial interpretation need to be revised. Reformulation is discussed in greater detail in Section 11.6.

## 3.4 How to Learn Data Science

Learning data science is much like learning a language or learning to play an instrument – you have to practice. Our advice based on mentoring many students and clients is to get started sooner rather than later and to accept that the code you'll write in the future will always be better than the code you'll write today. Also, many of the small details that separate a proficient data scientist from a novice can only be learned through practice as there are too many small details to learn them all in advance. So, starting today, do your best to write at least some code for all your projects. If a time deadline prevents you from completing the analysis in R, that's fine, but at least gain the experience of making an RStudio project and loading the data in R.[4] Then, as time allows, try to duplicate your analyses in R, being quick to search for solutions when you run into errors. Often simply copying and pasting your error into a search engine will be enough to find the solution to your problem. Moreover, searching for solutions is its own skill that also requires practice. Finally, if you are really stuck, reach out to a colleague (or even the authors of this book) for help.

## 3.5 Cautions: Don't Do That Everybody Does

We have all been in situations in which, for a given study, we edited the raw data files (e.g. removed respondents who were not present for the full study) and saved them using a different name. Some time later, as we need to get back to this study, or share the data with colleagues, finding the *correct* file quickly becomes a challenge that may end up being time consuming.

It seems clear that such way of working is not viable, and as mentioned earlier, raw data should never be edited. Instead, we prefer to run every data manipulation step (e.g. removing respondents) in R by also commenting why certain decisions are being made. This simplifies deeply the workflow, and in the future, you will be grateful when you will reopen this file and can easily find out what was done and why.

The same also applies for the analysis part. Documenting which analyses and which parameters were used ensures reproducible research. If at first, documenting your code may seem to be *a loss of time*, but it will pay back later when you will access your code again later in time, as decisions taken while you are writing your code may not be so clear anymore afterwards.

[4] We recommend following the instructions in Appendix 2 to get started.

Another important aspect is time: do not always go for the fastest or (what seems to be) easiest solution when coding. Instead, try to find the best possible balance between easy/fast coding and smart/efficient coding. For example, it may seem simpler to hard-code the names of the variables as they are in your data set than to read them from the file. However, this approach means that your code is restricted to that particular study or to any study that exactly fits that format. But as soon as there is a small change (e.g. a small difference in the naming of one of the variables), it will quickly cost you a lot of time to adapt the code to your new study.

Talking about efficiency, it is also advised to never use (through copy/paste for instance) the same lines of code more than twice. If you apply the same code multiple times in your analysis, or eventually across script files, consider better alternatives such as loops and/or functions. This point is of utmost importance as any small change in that piece of code (e.g. changing a parameter, fixing a bug, etc.) only needs to be done once to be applied everywhere. On the other hand, if you reproduce the same code multiple times, you need to ensure that you correctly modified each and every part that contains that code (and it is easy to skip some!)

Last but least, remember that coding is an endless process, as it can always be improved. So do not hesitate to go back to your own code and update it to make it more efficient, more flexible, more concise, etc. as you learn new things, or as new tools are being made available.

With these preliminaries completed, and with you (hopefully) sufficiently motivated, let's begin learning data science!

# 4

# *Data Manipulation*

This chapter aims in introducing the {tidyverse} as you'll learn how to manipulate and transform your data by exploring simple (e.g. renaming columns, sorting tables, relocating variables, etc.) to complex transformations (e.g. transposing parts of the table, combining/merging tables, etc.). Such transformations are done through simple examples that are relevant to Sensory and Consumer Science.

## 4.1 Why Manipulating Data?

In this chapter, many transformations proposed are not being saved. If you want to apply these changes to your data set, please visit Section 2.3.4. Moreover, some of the examples presented here emphasize the *how to*?, not the *why*?, and are not necessarily chosen to convey a scientific meaning.

Last but not least, most functions used are from the {tidyverse}. Let's start with loading this package:

```
library(tidyverse)
```

In sensory science, different data collection tools (e.g. different devices, software, methodologies, etc.) may provide the same data in different ways. Also, different statistical analyses may require having the data structured differently.

A simple example to illustrate this latter point is the analysis of liking data.

Let $C$ consumers provide their hedonic assessments on $P$ samples. To evaluate if samples have received different mean liking scores at the population level, an ANOVA is performed on a long thin table with 3 columns (consumer, sample, and the liking scores), where the combination of $C \times P$ is spread in rows (cf. Table 4.1).

DOI: 10.1201/9781003028611-4

**TABLE 4.1**

Example of a table for an analysis of variance

| Consumer | Product | Liking |
|---|---|---|
| C1 | P1 | 6 |
| C1 | P2 | 8 |
| ... | ... | ... |
| C1 | P | 5 |
| C2 | P1 | 8 |
| ... | ... | ... |
| c | p | L(c,p) |
| C | P | L(C,P) |

However, to assess whether consumers have the same preference patterns at the individual level, internal preference mapping or cluster analysis is performed. Both these analyses require as input a short and large table with the $P$ products in rows and the $C$ consumers in columns (cf. Table 4.2).

**TABLE 4.2**

Example of a table for internal preference mapping or clustering

| Product | C1 | C2 | ... | ... | C | |
|---|---|---|---|---|---|---|
| P1 | 6 | 8 | . | . | . | . |
| P2 | 8 | 4 | . | . | . | . |
| ... | ... | ... | . | . | . | . |
| p | 3 | 7 | . | L(p,c) | . | L(p,C) |
| ... | ... | ... | . | . | . | . |
| P | 5 | 6 | . | L(P,c) | . | L(P,C) |

Another example of data manipulation consists of summarizing data, by for instance computing the mean by product of the liking scores or by generating frequency tables (e.g. distribution of the liking scores by product). In this case, the transformation alters the data as the individual differences are lost. Ultimately, the output table is smaller as it would only contain $P$ rows and 2 columns (cf. Table 4.3):

**TABLE 4.3**

Example of mean table

| Product | Liking |
|---|---|
| P1 | 6.4 |
| P2 | 8.2 |
| ... | ... |
| p | 7.1 |
| ... | ... |
| P | 6.1 |

For these reasons, it is essential to learn to manipulate data and transition from one structure to another (when possible).

## 4.2 Tidying Data

Hadley Wickham (Wickham, 2014) defined *tidy data* as "data sets that are arranged such that each variable is a column and each observation (or case) is a row." Depending on the statistical unit to consider and the analyses to perform, data may need to be manipulated to be presented in a tidy form.

### 4.2.1 Simple Manipulations

We define here as *simple manipulations* any data transformations that can easily be performed in other software such as Excel (using copy-paste, sorting and filtering, creating a pivot table, etc.). However, we strongly recommend performing any sorts of transformation in R as this will reduce the risk of errors, typically be faster, more reliable, and will be reusable if you need to perform the same operations on similar data in the future (including updated versions of the current data set). Moreover, these operations will become easier and more natural for you to use as you get familiar with them. Most importantly, performing these transformations in R do not alter the original data, meaning that changes can be reverted (which is not possible when you alter the raw data directly).

#### *Handling Columns*

*Renaming Variables*

Before starting transforming any data, we need data. So let's start with importing the *biscuits_sensory_profile.xlsx* file.[1] For importing the data, the packages {`here`} and {`readxl`} are being used. Here, the data are being saved in the object called `sensory`:

```
library(here)
library(readxl)
file_path <- here("data", "biscuits_sensory_profile.xlsx")
sensory <- read_xlsx(file_path, sheet = "Data")
```

[1] For more details about the data set and/or data importation, please see Section 8.4.

The first simple transformation we consider consists of renaming one or multiple variables. This procedure can easily be done using the `rename()` function from {`dplyr`}. In each of the examples below, we use the `names()` function to show the column names (only) of the resulting data set.

```
sensory %>%
  names()
```

```
## [1] "Judge"
## [2] "Product"
## [3] "Shiny"
## [4] "External color intensity"
## [5] "Color evenness"
```

In `sensory`, let's rename `Judge` into `Panellist` and `Product` into `Sample` (here we apply transformations without saving the results, so the original data set remains unchanged).

To do so, we indicate in `rename()` that *new_name* is replacing *old_name* as following: `rename(newname = oldname)`. Additionally, we can apply multiple changes by simply separating them with a `,`:

```
sensory %>%
  rename(Panellist = Judge, Sample = Product) %>%
  names()
```

```
## [1] "Panellist"
## [2] "Sample"
## [3] "Shiny"
## [4] "External color intensity"
## [5] "Color evenness"
```

Alternatively, it is also possible to rename a column using its position:

```
sensory %>%
  rename(Consumer = 1, Biscuit = 2) %>%
  names()
```

```
## [1] "Consumer"
## [2] "Biscuit"
## [3] "Shiny"
## [4] "External color intensity"
## [5] "Color evenness"
```

If this procedure of renaming variables should be applied on many variables following a structured form (e.g. transforming names into snake_case, CamelCase, ..., see https://en.wikipedia.org/wiki/Letter_case\#Use_within_programming_languages for more information), the use of the {janitor} package comes handy thanks to its clean_names() function and the case parameter:

```
library(janitor)
sensory %>%
  clean_names(case = "snake") %>%
  names()
```

```
## [1] "judge"
## [2] "product"
## [3] "shiny"
## [4] "external_color_intensity"
## [5] "color_evenness"
```

Note that the {janitor} package offers many options, and although the transformation is performed here on all the variables, it is possible to apply it on certain variables only.

*Reorganizing Columns*

Another simple transformation consists in reorganizing the data set, either by reordering, adding, and/or removing columns.

For reordering columns, relocate() is being used. This function allows repositioning a (set of) variable(s) before or after another variable. By reusing the sensory data set, let's position all the variables starting with "Qty" between Product and Shiny. This can be specified in two different ways, either by positioning them after Product or before Shiny:

```
sensory %>%
  relocate(starts_with("Qty"), .after = Product) %>%
  names()
```

```
## [1] "Judge"
## [2] "Product"
## [3] "Qty of inclusions"
## [4] "Qty of inclusions in mouth"
## [5] "Shiny"
```

```
sensory %>%
  relocate(starts_with("Qty"), .before = Shiny) %>%
  names()
```

```
## [1] "Judge"
## [2] "Product"
## [3] "Qty of inclusions"
## [4] "Qty of inclusions in mouth"
## [5] "Shiny"
```

*Removing/Selecting Columns*

Another very important function regarding column transformation is the `select()` function from {dplyr} (see Section 2.2.1 for a justification of the particular writing `dplyr::select()`) which allows selecting a set of variables, by simply informing the variables that should be kept in the data. Let's limit ourselves in selecting `Judge`, `Product`, and `Shiny`:

```
sensory %>%
  dplyr::select(Judge, Product, Shiny)
```

```
## # A tibble: 99 x 3
##   Judge Product Shiny
##   <chr> <chr>   <dbl>
## 1 J01   P01      52.8
## 2 J01   P02      48.6
## 3 J01   P03      48
## 4 J01   P04      46.2
## # ... with 95 more rows
```

When a long series of variables should be kept in the same order, the use of the `:` is used.

Let's keep all the variables going from `Cereal flavor` to `Dairy flavor`:

```
sensory %>%
  dplyr::select(Judge, Product, `Cereal flavor`:`Dairy flavor`)
```

```
## # A tibble: 99 x 6
##   Judge Product `Cereal flavor` RawDo~1 Fatty~2 Dairy~3
##   <chr> <chr>             <dbl>   <dbl>   <dbl>   <dbl>
## 1 J01   P01                24.6    28.2    13.8       0
## 2 J01   P02                25.8    28.8     7.2       0
## 3 J01   P03                30      26.4     0         0
```

```
## 4 J01    P04                  16.2     28.2      0          0
## # ... with 95 more rows, and abbreviated variable
## #   names 1: 'RawDough flavor', 2: 'Fatty flavor',
## #   3: 'Dairy flavor'
```

However, when only one (or few) variable needs to be removed, it is easier to specify which one to remove rather than informing all the ones to keep. Such solution is then done using the - sign:

```
sensory %>%
  dplyr::select(-c(Shiny, Melting))
```

The selection process of variables can be further informed through functions such as `starts_with()`, `ends_with()`, and `contains()`, which all select variables that either starts, ends, or contains a certain character or sequence of character. To illustrate this, let's only keep the variables that starts with "Qty":

```
sensory %>%
  dplyr::select(starts_with("Qty"))
```

```
## # A tibble: 99 x 2
##    'Qty of inclusions' 'Qty of inclusions in mouth'
##                  <dbl>                        <dbl>
## 1                  9.6                         27.6
## 2                 10.8                         22.2
## 3                  7.8                         10.2
## 4                  0                           13.2
## # ... with 95 more rows
```

Rather than selecting variables based on their names, we can also select them based on their position (e.g. `dplyr::select(2:5)` to keep the variables that are at position 2 to 5).

Selection of variables can also be done using some *rules* thanks to the `where()` function. Let's consider the situation in which we only want to keep the variables that are nominal (or *character* in R), which automatically keeps `Judge` and `Product`:

```
sensory %>%
  dplyr::select(where(is.character))
```

```
## # A tibble: 99 x 2
##   Judge Product
##   <chr> <chr>
## 1 J01   P01
## 2 J01   P02
## 3 J01   P03
## 4 J01   P04
## # ... with 95 more rows
```

`dplyr::select()` is a very powerful function that facilitates the selection of complex variables through very intuitive functions. Ultimately, it can also be used to `relocate()` and even `rename()` variables, as shown in the example below:

```
sensory %>%
  dplyr::select(Panellist = Judge, Sample = Product,
                Shiny:Thickness, -contains("olor"))
```

More examples illustrating the use of `dplyr::select()` are provided throughout the book. In particular, in the next Section "Handling Rows", another important function called `across()` will be introduced. This function allows applying the same transformation to multiple columns, allowing you to use select() semantics inside functions such as summarise() and mutate().

*Creating Columns*

In some cases, new variables need to be created from existing ones. Examples of such situations include taking the quadratic term of a sensory attribute to test for curvature or simply considering a new variables as the sum or the subtraction between two (or more) others. Such creation of a variable is processed through the `mutate()` function from the {dplyr} package. This function takes as inputs the name of the variable to create and the *formula* that defines that variable.

Let's create two new variables, one called `Shiny2` which corresponds to `Shiny` squared up and another one called `StiMelt` which corresponds to `Sticky` + `Melting`. Since we only use these three variables, let's first reduce the data set to these three variables with `select()` to improve readability:

```
sensory %>%
  dplyr::select(Shiny, Sticky, Melting) %>%
  mutate(Shiny2 = Shiny^2,
         StiMelt = Sticky + Melting)
```

```
## # A tibble: 99 x 5
##    Shiny Sticky Melting Shiny2 StiMelt
##    <dbl>  <dbl>   <dbl>  <dbl>   <dbl>
## 1  52.8   37.2    33.6  2788.    70.8
## 2  48.6   35.4    36    2362.    71.4
## 3  48     37.2     8.4  2304     45.6
## 4  46.2   21.6    34.2  2134.    55.8
## # ... with 95 more rows
```

---

If you want to transform a variable, say by changing its type or rewriting its content, you can use `mutate()` and assign to the new variable the same name as the original one. This will overwrite the existing column with the new one. To illustrate this, let's transform `Product` from upper case to lower case only. This can be done by mutating `Product` into the lowercase version of `Product` (`tolower(Product)`): `sensory %>% mutate(Product = tolower(Product))`

`mutate()` being one of the most important function from the {`dplyr`} package; more examples of its use are presented throughout this book.

---

Since performing mathematical computations on nonnumerical columns is not possible, conditions can easily be added through `mutate()` combined with `across()`. Let's imagine we want to round all variables to 0 decimal, which can only be applied to numerical variables. To do so, we `mutate()` `across()` all variables that are considered `as.numeric()` (using `where()`):

```
sensory %>%
  mutate(across(where(is.numeric), round, digits = 0))
```

```
## i In argument: 'across(where(is.numeric), round, digits = 0)'.
## Caused by warning:
## ! The '...' argument of 'across()' is deprecated as of dplyr 1.1.0.
## Supply arguments directly to '.fns' through an anonymous function instead.
##
##   # Previously
##   across(a:b, mean, na.rm = TRUE)
##
##   # Now
##   across(a:b, \(x) mean(x, na.rm = TRUE))
```

```
## # A tibble: 99 x 34
##   Judge Product Shiny Externa~1 Color~2 Qty o~3 Surfa~4
##   <chr> <chr>   <dbl>     <dbl>   <dbl>   <dbl>   <dbl>
## 1 J01   P01        53        30      23      10      23
## 2 J01   P02        49        30      13      11      13
## 3 J01   P03        48        46      17       8      14
```

```
## 4 J01    P04           46           46          38          0          49
## # ... with 95 more rows, 27 more variables:
## #   `Print quality` <dbl>, Thickness <dbl>,
## #   `Color contrast` <dbl>,
## #   `Overall odor intensity` <dbl>,
## #   `Fatty odor` <dbl>, `Roasted odor` <dbl>,
## #   `Cereal flavor` <dbl>, `RawDough flavor` <dbl>,
## #   `Fatty flavor` <dbl>, `Dairy flavor` <dbl>, ...
```

In case only a selection of numerical variables should be rounded, we could also replace `where(is.numeric)` by a vector (using `c()`) with the names of the variables to round.

```
sensory %>%
  dplyr::select(Shiny, Sticky, Melting) %>%
  mutate(across(c("Shiny", "Sticky"), round, digits = 0))
```

*Merging and Separating Columns*

It can happen that some columns of a data set contain information (strings) that cover different types of information. For instance, we could imagine coding the name of our panelists as *FirstName_LastName* or *Gender_Name*, and we would want to separate them into two columns to make the distinction between the different information (i.e. *FirstName* and *LastName*, or *Gender* and *Name* respectively). In other situations, we may want to merge information present in multiple columns into one.

For illustration, let's consider the information stored in the *Product Info* sheet from *biscuits_sensory_profile.xlsx*. This table includes information regarding the biscuits, and more precisely their Protein and Fiber content (Low or High).

After importing the data, let's merge these two columns so that both information is stored in one column called `ProtFib`. To do so, `unite()` (from {`tidyr`}) is used. This function takes as first element the name of the new variables, followed by all the columns to *unite*, and by providing the separation key to use between these elements (here -):

```
prod_info <- read_xlsx(file_path, sheet = "Product Info") %>%
  unite(ProtFib, Protein, Fiber, sep = "-")
```

```
## # A tibble: 11 x 3
##   Product ProtFib  Type
##   <chr>   <chr>    <chr>
## 1 P01     Low-Low  Trial
```

```
## 2 P02     Low-High  Trial
## 3 P03     High-High Trial
## 4 P04     High-High Trial
## # ... with 7 more rows
```

By default, `unite()` removes from the data set the individual variables that have been merged. To keep these original variables, the parameter `remove = FALSE` can be used.

---

Although it is not relevant for combining columns, it is interesting to mention an additional package that can be used to combine elements together. This package is called {`glue`} and provides interesting alternatives to the usual `paste()` and `paste0()` functions.

---

To reverse the changes (saved here in `prod_info`) and to separate a column into different variables, `separate()` (from {`tidyr`}) is used. Similarly to `unite()`, `separate()` takes as first parameter the name of the variable to split, followed by the names for the different segments generated, and of course the separator defined by `sep`.

In our example, this would be done as following:

```
prod_info %>%
  separate(ProtFib, c("Protein", "Fiber"), sep = "-")
```

```
## # A tibble: 11 x 4
##   Product Protein Fiber Type
##   <chr>   <chr>   <chr> <chr>
## 1 P01     Low     Low   Trial
## 2 P02     Low     High  Trial
## 3 P03     High    High  Trial
## 4 P04     High    High  Trial
## # ... with 7 more rows
```

*Conditions*

In some cases, the new column to create depend directly on the value(s) of one or more columns present in the data. An example of such situations consists of categorizing a continuous variable into groups by converting the age (in year) of the participants into age groups. For such simple examples, some pre-existing functions (e.g. `cut()` in this situation) can be used. However, in other situations, predefined functions do not exist and the transformation should be done *manually* using conditions.

Let's illustrate this by converting the `Shiny` variable (from `sensory`) from numeric to classes. Since the scale used is a 60pt scale, let's start by creating a class called *Low* if the score is lower than 30 and *High* otherwise.

Here, predefined functions (e.g. `cut()`) are not being used intentionally as a manual transformation is preferred. Instead, `mutate()` is associated to `ifelse()`, which works as following: `ifelse(condition, results if condition is TRUE, results if condition is not TRUE)`

```
sensory %>%
  dplyr::select(Shiny) %>%
  mutate(ShinyGroup = ifelse(Shiny < 30, "Low", "High"))
```

```
## # A tibble: 99 x 2
##   Shiny ShinyGroup
##   <dbl> <chr>
## 1  52.8 High
## 2  48.6 High
## 3  48   High
## 4  46.2 High
## # ... with 95 more rows
```

Let's imagine the same variable should now be split into three levels: *Low*, *Medium*, and *High*. Such additional group could be obtained by adding an `ifelse()` condition within the existing `ifelse()` condition (we use 48 instead of 40 for the upper limit to `Medium` so that results are displayed on screen):

```
sensory %>%
  dplyr::select(Shiny) %>%
  mutate(ShinyGroup = ifelse(Shiny < 20, "Low",
                       ifelse(Shiny < 48, "Medium", "High")))
```

```
## # A tibble: 99 x 2
##   Shiny ShinyGroup
##   <dbl> <chr>
## 1  52.8 High
## 2  48.6 High
## 3  48   High
## 4  46.2 Medium
## # ... with 95 more rows
```

Since there are only three conditions in total here, only two entangled `ifelse()` are required. This makes the code still manageable. However, in more complex situations (say 10 different conditions are required), such solution quickly becomes tedious to read, track, and debug if errors are being made. Instead, the use of an alternative function called `case_when()` is preferred. In the previous case, the same conditions would be written as follows:

```
sensory %>%
  dplyr::select(Shiny) %>%
  mutate(ShinyGroup = case_when(
    Shiny < 20 ~ "Low",
    between(Shiny, 20, 48) ~ "Medium",
    Shiny > 40 ~ "High"))
```

```
## # A tibble: 99 x 2
##   Shiny ShinyGroup
##   <dbl> <chr>
## 1  52.8 High
## 2  48.6 High
## 3  48   Medium
## 4  46.2 Medium
## # ... with 95 more rows
```

This provides the same results as previously, except for the exact value 48 which was assigned as *High* in the `ifelse()` example and to *Medium* in the `case_when()` example. This is due to the way `between()`[2] considers its borders.

## Handling Rows

After manipulating columns, the next logical step is to manipulate rows. Such operations include four aspects:

1. Rearranging the rows in a logical way;
2. Selecting certain rows;
3. Filtering entries based on given variables;
4. Splitting the data into subgroups based on the entries of a given variable.

### *Rearranging Rows*

The first step of rearranging rows is done through the `arrange()` function from the {`dplyr`} package. This function allows sorting the data in the ascending order.[3] To arrange them in a descending order, the function `desc()` is then required.

Let's rearrange the data by `Judge` and `Product`, with `Judge` being sorted in an ascending order whereas `Product` is being sorted in a descending order:

[2] By default, `between(x, value1, value2)` considers value1 <= x <= value2.

[3] For numerical order, this is simply rearranging the values from the lowest to the highest. For strings, the entries are then sorted alphabetically unless the variable is a factor in which case the order of the levels for that factor is being used.

```
sensory %>%
  arrange(Judge, desc(Product))
```

```
## # A tibble: 99 x 34
##   Judge Product Shiny Externa~1 Color~2 Qty o~3 Surfa~4
##   <chr> <chr>   <dbl>     <dbl>   <dbl>   <dbl>   <dbl>
## 1 J01   POpt      4.8      33.6    15.6    32.4    13.8
## 2 J01   P10      53.4      36.6    11.4    18      10.8
## 3 J01   P09       0        42.6    18      21      36
## 4 J01   P08       0        51.6    48.6    23.4    18
## # ... with 95 more rows, 27 more variables:
## #   'Print quality' <dbl>, Thickness <dbl>,
## #   'Color contrast' <dbl>,
## #   'Overall odor intensity' <dbl>,
## #   'Fatty odor' <dbl>, 'Roasted odor' <dbl>,
## #   'Cereal flavor' <dbl>, 'RawDough flavor' <dbl>,
## #   'Fatty flavor' <dbl>, 'Dairy flavor' <dbl>, ...
```

*Selecting Rows*

The next step is to select a subset of the data by keeping certain rows only. If the position of the rows to keep is known, this information can be used directly using the `slice()` function. Let's select from `sensory` all the data that is related to `P01`. A quick look at the data informs us that it corresponds to rows 1 to 89, with a step of 11:

```
sensory %>%
  slice(seq(1, 89, 11))
```

```
## # A tibble: 9 x 34
##   Judge Product Shiny Externa~1 Color~2 Qty o~3 Surfa~4
##   <chr> <chr>   <dbl>     <dbl>   <dbl>   <dbl>   <dbl>
## 1 J01   P01      52.8      30      22.8     9.6    22.8
## 2 J02   P01      44.4      34.2    14.4    18.6    43.2
## 3 J03   P01      40.2      23.4     9       7.8    49.8
## 4 J04   P01      37.8      26.4    15      23.4    15.6
## # ... with 5 more rows, 27 more variables:
## #   'Print quality' <dbl>, Thickness <dbl>,
## #   'Color contrast' <dbl>,
## #   'Overall odor intensity' <dbl>,
## #   'Fatty odor' <dbl>, 'Roasted odor' <dbl>,
## #   'Cereal flavor' <dbl>, 'RawDough flavor' <dbl>,
## #   'Fatty flavor' <dbl>, 'Dairy flavor' <dbl>, ...
```

This is a manual way to select data. However, this procedure may generate an erroneous subset as soon as the row order in the data changes.

To avoid mistakes, a more *stable* procedure of filtering data is proposed in the next paragraph.

*Filtering Data*

To define the subset of data, the `filter()` function is being used. This function requires providing an argument that is expressed as a *test*, meaning that the outcome should either be TRUE (keep the value) or FALSE (discard the value) when the condition is verified or not, respectively. In R, this is expressed by the double "=" sign `==`.

Let's filter the data to only keep the data related to sample `P02`:

```
sensory %>%
  filter(Product == "P02")
```

```
## # A tibble: 9 x 34
##   Judge Product Shiny Externa~1 Color~2 Qty o~3 Surfa~4
##   <chr> <chr>   <dbl>     <dbl>   <dbl>   <dbl>   <dbl>
## 1 J01   P02      48.6      30      13.2    10.8    13.2
## 2 J02   P02      39.6      32.4    18      19.8    25.2
## 3 J03   P02      39        18.6    13.2     9      28.8
## 4 J04   P02      39.6      41.4    33      25.2    10.2
## # ... with 5 more rows, 27 more variables:
## #   `Print quality` <dbl>, Thickness <dbl>,
## #   `Color contrast` <dbl>,
## #   `Overall odor intensity` <dbl>,
## #   `Fatty odor` <dbl>, `Roasted odor` <dbl>,
## #   `Cereal flavor` <dbl>, `RawDough flavor` <dbl>,
## #   `Fatty flavor` <dbl>, `Dairy flavor` <dbl>, ...
```

Other relevant test characters are the following:

- `!Product == "P02"` or `Product != "P02"` means different from, and will keep all samples except `P02`;
- `%in% my_vector` keeps any value included within the vector `my_vector` (e.g. `Product %in% c("P01","P02","P03")` keeps all data from `P01`, `P02`, and `P03`).

In some cases, the tests to perform are more complex as they require multiple conditions. There are two forms of conditions:

- `&` (read *and*) is multiplicative, meaning that all the conditions need to be true (`Product == "P02" & Shiny > 40`);
- `|` (read *or*) is additive, meaning that only one of the conditions needs to be true (`Product == "P03" | Shiny > 40`).

This system of condition is particularly useful when you have missing values as you could remove all the rows that contain missing values for a given variable. Since we do not have missing values here, let's create some by replacing all the evaluations for `Shiny` that are larger than 40 by missing values. This is done here using `ifelse()`, which takes three arguments (in this order): the test to perform (here `Shiny > 40`), the instruction if the test passes (here replace with `NA`), and the instruction if the test doesn't pass (here keep the value stored in `Shiny`).

```
sensory_na <- sensory %>%
  dplyr::select(Judge, Product, Shiny) %>%
  mutate(Shiny = ifelse(Shiny > 40, NA, Shiny))
```

In a second step, we filter out all missing values from `Shiny`. In practice, this is done by keeping all the values that are not missing:

```
sensory_na %>%
  filter(!is.na(Shiny))
```

This procedure removed 20 rows since the original table had 99 rows and 3 columns, whereas the filtered table only has 79 rows and 3 columns.

*Splitting Data*

After filtering data, the next logical step is to split data into subsets based on a given variable (e.g. by gender). For such purpose, one could consider using `filter()` by applying it to each subgroup. In a previous example, this is what we have done when we filtered data for sample `P02` only. Of course, the same procedure can be performed until all the other subgroups are created. However, this solution becomes tedious as the number of samples increases. Instead, we prefer to use `split()` which takes as arguments the data and the column to split from:

```
sensory %>%
  split(.$Product)
```

This function creates a list of $n$ elements ($n$ being the number of samples), each element corresponding to the data related to one sample. Such list can then be used in automated analyses by performing on each sub-data through the `map()` function, as it will be illustrated in Section 10.

### 4.2.2 Reshaping Data

Reshaping the data itself is done through pivoting which allows transitioning from a long and thin table to a short and wide table and vice versa.

To illustrate this, let's start with a fictive example in which we have three consumers providing their liking scores for two products. This table corresponds to a long and thin format (cf. Table 4.4):

**TABLE 4.4**

Fictive example shaped in a long thin formats

| Consumer | Product | Liking |
|---|---|---|
| C1 | P1 | 8 |
| C1 | P2 | 5 |
| C2 | P1 | 9 |
| C2 | P2 | 6 |
| C3 | P1 | 7 |
| C3 | P2 | 4 |

Let's imagine we need to restructure the data where products are displayed in rows and consumers in columns. This corresponds to the short and wide format (cf. Table 4.5):

**TABLE 4.5**

Fictive example restructured in a short wide format

| Product | C1 | C2 | C3 |
|---|---|---|---|
| P1 | 8 | 9 | 7 |
| P2 | 5 | 6 | 4 |

As we will see in the following section, it is very easy to transition from one version to another thanks to `pivot_longer()` and `pivot_wider()`, both being functions from {`tidyr`}.

#### *Pivoting Longer*

Currently, our `sensory` data table is a table in which we have as many rows as Judge x Product, the different attributes being spread across multiple columns. However, in certain situations, it is relevant to have all the attributes stacked vertically, meaning that the table will have Judge x Product x Attributes rows. Such simple transformation can be done with `pivot_longer()` which takes as inputs the attributes to pivot, the name of the variables that will contain these names (`names_to`), and the name of the column that will contain their entries (`values_to`)

---

With `pivot_longer()` and any other function that requires selecting variables, it is often easier to deselect variables that we do not want to include rather than selecting all the variables of interest. Throughout the book, both solutions are being considered.

---

```
sensory %>%
  pivot_longer(Shiny:Melting,
               names_to = "Attribute", values_to = "Score")
```

```
## # A tibble: 3,168 x 4
##   Judge Product Attribute                 Score
##   <chr> <chr>   <chr>                     <dbl>
## 1 J01   P01     Shiny                      52.8
## 2 J01   P01     External color intensity  30
## 3 J01   P01     Color evenness             22.8
## 4 J01   P01     Qty of inclusions           9.6
## # ... with 3,164 more rows
```

This transformation converts a table of 99 rows and 34 columns into a table containing 3168 ($99 \times 32$) rows and 4 columns. In the pivoted table, the names of the variables (stored here in `Attribute`) are in the same order as presented in the original table.

In case the attribute names are following a standard structure, say "attribute_name modality" as is the case in `sensory` for some attributes, an additional parameter of `pivot_longer()` becomes handy as it can split the `Attribute` variable just created into say `Attribute` and `Modality`.

To illustrate this, let's reduce `sensory` to `Judge`, `Product`, and all the variables that end with odor or flavor (all other variables being discarded). After pivoting the subset of columns, we automatically split the attribute names into `Attribute` and `Modality` by informing the separator between names (here, a space):

```
sensory %>%
  dplyr::select(Judge, Product,
                ends_with("odor"), ends_with("flavor")) %>%
  pivot_longer(-c(Judge, Product),
               names_to = c("Attribute", "Modality"),
               values_to = "Score", names_sep = " ")
```

```
## # A tibble: 693 x 5
##   Judge Product Attribute Modality Score
```

```
##    <chr> <chr>  <chr>     <chr>     <dbl>
## 1 J01   P01    Fatty     odor        6.6
## 2 J01   P01    Roasted   odor       15.6
## 3 J01   P01    Cereal    flavor     24.6
## 4 J01   P01    RawDough  flavor     28.2
## # ... with 689 more rows
```

This parameter combines both the power of `pivot_longer()` and `separate()` in one unique process.

Note that more complex transformations through the use of regular expressions (and the `names_pattern` option) can be considered. More information on regular expression is provided in Section 10.2.2.

---

It can happen that with `pivot_longer()`, further transformation performed on the long and thin table may not maintain their original order (usually, the names are reordered alphabetically). If you don't want such reordering to happen because it would impact the desired outcome, there is a simple workaround that ensures that the order is kept. The solution simply consists in transforming the newly created variable as a factor which takes as levels the order of the elements as they were in the original data (use `fct_inorder()` to maintain the order as shown in the data). An example is shown in the following code:

```
sensory %>%
  dplyr::select(Judge, Product, Shiny, Salty, Bitter, Light) %>%
  pivot_longer(-c("Judge", "Product"),
               names_to = "Variable", values_to = "Score") %>%
  mutate(Variable = fct_inorder(Variable))
```

Other examples using this trick will be used throughout this book (e.g. see Section 5.4.5)

---

As an alternative to `pivot_longer()`, the package called {reshape2} provides a function called `melt()` which pivots automatically the entire set of numerical variables, the qualitative variables being considered as *id variables*. If performed on a matrix with row names, the new table will have two columns containing the row and column names.

```
library(reshape2)
melt(sensory)
```

### *Pivoting Wider*

The complementary/opposite function to `pivot_longer()` is `pivot_wider()`. This function pivots data horizontally, hence reducing the number of rows and increasing the number of columns. In this case, the two main parameters to provide is which column will provide the new columns to create (`name_from`), and what are the values to use to fill this table (`values_from`).

From the previous example, we could set `names_from = Attribute` and `values_from = Score` to return to the original format of `sensory`. However, let's reduce the data set to `Product`, `Judge`, and `Shiny` only, and let's pivot the `Judge` and `Shiny` columns:

```
sensory %>%
  dplyr::select(Judge, Product, Shiny) %>%
  pivot_wider(names_from = Judge, values_from = Shiny)
```

```
## # A tibble: 11 x 10
##   Product   J01   J02   J03   J04   J05   J06   J07
##   <chr>   <dbl> <dbl> <dbl> <dbl> <dbl> <dbl> <dbl>
## 1 P01      52.8  44.4  40.2  37.8  43.8  43.2  44.4
## 2 P02      48.6  39.6  39    39.6  33.6  38.4  25.8
## 3 P03      48    36    35.4  15    37.2  33    16.2
## 4 P04      46.2  36    48    38.4  47.4  37.8  27
## # ... with 7 more rows, and 2 more variables:
## #   J08 <dbl>, J09 <dbl>
```

This procedure creates a table with as many rows as there are products, and as many columns as there are panelists (+1 since the product information is also in a column).

These procedures are particularly useful in consumer studies, since `pivot_longer()` and `pivot_wider()` allow restructuring the data for analysis such as ANOVA (`pivot_longer()` output) and preference mapping or clustering (`pivot_wider()` structure).

It is important to notice that the `pivot_wider()` format potentially contains more data. Let's imagine the sensory test was performed following an incomplete design, meaning that each panelist did not evaluate all the samples. Although the long and thin structure would not show missing values (the entire rows without data being removed), the shorter and larger version would contain missing values for the products that panelists did not evaluate. If the user wants to automatically replace these missing values with a fixed value, say, it is possible through the parameter `values_fill` (e.g. `values_fill=0` would replace each missing value with a 0). Additionally, after pivoting the data, if multiple entries exist for a combination row-column, `pivot_wider()` will return a list of elements. In the Section 4.2.3, an example illustrating such situation and how to handle it will be presented.

### 4.2.3 Transformation That Alters the Data

In some cases, the final table to generate requires altering the data, by (say) computing the mean across multiple values or counting the number of occurrences of factor levels for instance. In other words, we summarize the information, which also tend to reduce the size of the table. It is hence no surprise that the function used for such data reduction is called `summarise()` (or `summarize()`, both notation work) and belongs to the {`dplyr`} package.

#### *Introduction to Summary Statistics*

In practice, `summarise()` applies a function (whether it is the `mean()` or a simple count using `n()` for instance) on a set of values. Let's compute the mean on all numerical variables of `sensory`:

```
sensory %>%
  summarise(across(where(is.numeric), mean))
```

```
## # A tibble: 1 x 32
##   Shiny Exter~1 Color~2 Qty o~3 Surfa~4 Print~5 Thick~6
##   <dbl>   <dbl>   <dbl>   <dbl>   <dbl>   <dbl>   <dbl>
## 1  23.9    33.7    28.2    20.6    23.3    40.7    25.5
## # ... with 25 more variables: 'Color contrast' <dbl>,
## #   'Overall odor intensity' <dbl>,
## #   'Fatty odor' <dbl>, 'Roasted odor' <dbl>,
## #   'Cereal flavor' <dbl>, 'RawDough flavor' <dbl>,
## #   'Fatty flavor' <dbl>, 'Dairy flavor' <dbl>,
## #   'Roasted flavor' <dbl>,
## #   'Overall flavor persistence' <dbl>, ...
```

As can be seen, the grand mean is computed for each attribute. It can also be noticed that all the other variables that were not involved have been removed (e.g. `Judge` and `Product` as they are not numerical variables).

If multiple functions should be applied, we could perform all the transformation simultaneously as following:

```
sensory %>%
  summarise(across(where(is.numeric), list(min = min, max = max)))
```

```
## # A tibble: 1 x 64
##   Shiny_min Shiny_max Externa~1 Exter~2 Color~3 Color~4
##       <dbl>     <dbl>     <dbl>   <dbl>   <dbl>   <dbl>
## 1         0        54       6.6    55.2     6.6    53.4
## # ... with 58 more variables:
```

```
## #   `Qty of inclusions_min` <dbl>,
## #   `Qty of inclusions_max` <dbl>,
## #   `Surface defects_min` <dbl>,
## #   `Surface defects_max` <dbl>,
## #   `Print quality_min` <dbl>,
## #   `Print quality_max` <dbl>, ...
```

In this example, each attribute is duplicated with `_min` and `_max` to provide the minimum and maximum values for each attribute.

---

It would be a good exercise to restructure this table using `pivot_longer()` with `names_sep` followed by `pivot_wider()` to build a new table that shows for each attribute (in rows) the minimum and the maximum in two different columns.

---

By following the same principles, many other functions can be performed, whether they are built-in R or created by the user.

Here is a recommendation of interesting descriptive functions to consider with `summarise()`:

- `mean()`, `median()` (or more generally `quantile()`) for the mean and median (or any other quantile);
- `sd()` and `var()` for the standard deviation and the variance;
- `min()`, `max()`, `range()` (provides both the min and max) or `diff(range())` (for the difference between min and max);
- `n()` and `sum()` for the number of counts and the sum, respectively.

### *Introduction to Grouping*

It can appear that the interest is not in the grand mean, but in the mean per product (say), or per product and panelist (for test with duplicates). In such cases, `summarize()` should aggregate set of values per product or per product and panelist, respectively. Such information can be passed on through `group_by()`.[4]

```
sensory %>%
  group_by(Product) %>%
  summarise(across(where(is.numeric), mean)) %>%
  ungroup()
```

[4] We strongly recommend you to `ungroup()` blocks of code that includes `group_by()` once the computations are done to avoid any unexpected results. Otherwise, further computations may be done on the groups when it should be performed on the full data.

```
## # A tibble: 11 x 33
##    Product Shiny Exter~1 Color~2 Qty o~3 Surfa~4 Print~5
##    <chr>   <dbl>   <dbl>   <dbl>   <dbl>   <dbl>   <dbl>
## 1 P01       41.9    26.2    15.8    15.5    27.7    37.7
## 2 P02       39.1    29.5    20.3    14      20.5    39.5
## 3 P03       30.5    43.6    30.7    17.6    18.6    43.3
## 4 P04       42.6    43.3    37.7    15.1    32.8    30.3
## # ... with 7 more rows, 26 more variables:
## #   Thickness <dbl>, 'Color contrast' <dbl>,
## #   'Overall odor intensity' <dbl>,
## #   'Fatty odor' <dbl>, 'Roasted odor' <dbl>,
## #   'Cereal flavor' <dbl>, 'RawDough flavor' <dbl>,
## #   'Fatty flavor' <dbl>, 'Dairy flavor' <dbl>,
## #   'Roasted flavor' <dbl>, ...
```

This procedure creates a tibble with 11 rows (product) and 33 columns (32 sensory attributes + 1 column including the product information) which contains the mean per attribute for each sample, also known as the sensory profiles of the products.

---

In some cases, the data should not be aggregated across rows, but by rows. It is then important to specify that each computation should be done per row by using `rowwise()` prior to performing the transformation. For instance, if we would want to extract the minimum between `Shiny`, `Salty`, and `Bitter`, we could write the following code:

```
sensory %>%
  dplyr::select(Judge, Product, Shiny, Salty, Bitter)%>%
  rowwise()%>%
  mutate(Min = min(Shiny, Salty, Bitter))
```

---

### *Illustrations of Data Manipulation*

Let's review the different transformations presented here by generating the sensory profiles of the samples through different approaches.[5]

In the previous example, we've seen how to obtain the sensory profile using `summarise()` `across()` all numerical variables. In case a selection of the attributes should have been done, we could use the same process by simply informing which attributes to transform:

[5] It is important to realize that any data manipulation challenge can be tackled in many different ways, so don't be afraid to think out of the box when solving them.

```
sensory %>%
  group_by(Product) %>%
  summarise(across(Shiny:Melting, mean)) %>%
  ungroup()
```

The list of attributes to include can also be stored in an external vector:

```
sensory_attr <- colnames(sensory)[5:ncol(sensory)]
sensory %>%
  group_by(Product) %>%
  summarise(across(all_of(sensory_attr), mean)) %>%
  ungroup()
```

A different approach consists of combining `summarise()` to `pivot_longer()` and `pivot_wider()`. This process requires summarizing only one column by Product and Attribute:

```
sensory %>%
  pivot_longer(Shiny:Melting,
               names_to = "Attribute", values_to = "Scores") %>%
  mutate(Attribute = fct_inorder(Attribute)) %>%
  group_by(Product, Attribute) %>%
  summarise(Scores = mean(Scores)) %>%
  pivot_wider(names_from = Attribute, values_from = Scores) %>%
  ungroup()
```

Here, we transformed `Attribute` into a factor using `fct_inorder()` to ensure that the double pivoting procedure maintains the original order. Without this line of code, the final table would have the columns reordered alphabetically.

---

As you can see, R provides the following message: *`summarise()` has grouped output by 'Product'. You can override using the `.groups` argument.* This message is just informative, and can be hidden by adding the following code at the start of your script: `options(dplyr.summarise.inform = FALSE)`.

---

What would happen if we would omit to `summarise()` the data in between the two pivoting functions? In that case, we also remove the column `Judge` since the means should be computed across panelists...

```
sensory %>%
  pivot_longer(Shiny:Melting,
               names_to = "Attribute", values_to = "Scores") %>%
  dplyr::select(-Judge) %>%
  pivot_wider(names_from = Attribute, values_from = Scores)
```

As can be seen, each variable is of type list in which each cell contains `dbl [9]`: This corresponds to the scores provided by the nine panelists to that product and that attribute. Since we would ultimately want the mean of these nine values to generate the sensory profiles, a solution comes directly within `pivot_wider()` through the parameter `values_fn` which applies the function provided here on each set of values:

```
sensory %>%
  pivot_longer(Shiny:Melting,
               names_to = "Attribute", values_to = "Scores") %>%
  dplyr::select(-Judge) %>%
  pivot_wider(names_from = Attribute, values_from = Scores,
              values_fn = mean)
```

Through this simple example, we've seen that the same results can be obtained through different ways. It is important to keep this in mind as you may find the solution to your own challenges by simply considering different paths.

### 4.2.4 Combining Data from Different Sources

It often happens that the data to analyze is stored in different files, and need to be combined or merged. Depending on the situations, different solutions are required.

#### *Binding Vertically*

Let's start with a simple example where the tables match in terms of variables and should be combined vertically. To illustrate this situation, the data stored in the file *excel-scrap.xlsx* is used. This file contains a fictive example in which 12 assessors evaluated 2 samples on 3 attributes in triplicate, with each replication being stored in a different sheet.

The goal here is to read the data stored in the different sheets and to combine them vertically in one unique file for further analysis. Let's start with importing the data:

```
path <- file.path("data", "excel_scrap.xlsx")
session1 <- read_xlsx(path, sheet = 1)
session2 <- read_xlsx(path, sheet = 2)
session3 <- read_xlsx(path, sheet = 3)
```

To combine the tables vertically, we could use the basic R function `rbind()`. However, we prefer to use `bind_rows()` from {`dplyr`} since it better controls for the columns by ensuring that the order is respected. Moreover, if one of the tables contains a variable that the other don't, this variable will be kept and filled in with missing values when the information is missing. Additionally, `bind_rows()` allows keeping track of the *origin* of the data through the parameter `.id`. This is of particular interest in this example since a new `Session` column can be created (and used) to distinguish between tables. This process is used to avoid losing such useful information, especially since it is not directly available within the data: If it were, the parameter `.id` could have been ignored.

This solution works fine, especially since there were only three files to combine. Ultimately, we would prefer to automate the reading process so that all the files are directly imported and combined. This more efficient solution is presented in Section 8.4.3.

### Binding Horizontally

In other cases, the tables to combine contain different information (variables) on the same entities (rows) and the tables should be merged horizontally. To do so, a first solution consists in using the functions `cbind()` ({`base`}) and/or `bind_cols()` ({`dplyr`}). However, some of these functions require that the tables to combine must already have the rows in the exact same order (no check is being done) and must be of the same size.

If that is not the case, merging tables should be done using `merge()` ({`base`}) or preferably through the different `*_join()` functions from ({`dplyr`}). For illustration, let's consider these two tables to merge (Tables 4.6 and 4.7):

**TABLE 4.6**

Tables used to illustrate the different mergeing options

| Product | Var1 |
|---|---|
| A | 1 |
| B | 2 |
| C | 3 |

**TABLE 4.7**

Tables used to illustrate the different mergeing options

| Product | Var2 |
|---|---|
| A | 1 |
| B | 2 |
| D | 4 |

Depending on the *merging degree* to consider between tables X and Y, there are four different `*_join()` versions to consider:

- `full_join()` (illustrated in Table 4.8) keeps all the cases from X and Y regardless whether they are present in the other table or not (in case they are not present, missing values are introduced) [it corresponds to `merge(all=TRUE)`];

**TABLE 4.8**

Illustration of the 'full_join()' option

| Product | Var1 | Var2 |
|---|---|---|
| A | 1 | 1 |
| B | 2 | 2 |
| C | 3 | |
| D | | 4 |

- `inner_join()` (illustrated in Table 4.9) only keeps the common cases, that is, cases that are present in both X and Y [corresponds to `merge(all=FALSE)`];

**TABLE 4.9**

Illustration of the 'inner_join()' option

| Product | Var1 | Var2 |
|---|---|---|
| A | 1 | 1 |
| B | 2 | 2 |

- `left_join()` (illustrated in Table 4.10) keeps all the cases from X [corresponds to `merge(all.x=TRUE, all.y=FALSE)`];

**TABLE 4.10**

Illustration of the 'left_join()' option

| Product | Var1 | Var2 |
|---|---|---|
| A | 1 | 1 |
| B | 2 | 2 |
| C | 3 | |

- `right_join()` (illustrated in Table 4.11) keeps all the cases from Y [corresponds to `merge(all.x=FALSE, all.y=TRUE)`];

**TABLE 4.11**

Illustration of the 'right_join()' option

| Product | Var1 | Var2 |
|---|---|---|
| A | 1 | 1 |
| B | 2 | 2 |
| D | | 4 |

- `anti_join()` (illustrated in Table 4.12) only keeps the elements from X that are not present in Y (this is particularly useful if you have a tibble Y of elements that you would like to remove from X).

**TABLE 4.12**

Illustration of the 'anti_join()' option

| Product | Var1 |
|---|---|
| C | 3 |

The merging procedure requires the users to provide a *key*, that is, a (set of) variable(s) used to combine the tables. For each unique element defined by the key, a line is being created. When needed, rows of a table are being duplicated. Within the different `*_join()` functions, the key is informed by the `by` parameter, which may contain one or more variables with the same or different names.

For illustration, let's use the data set called *biscuits_consumer_test.xlsx*, which contains three tabs:

```
file_path <- here("data", "biscuits_consumer_test.xlsx")
excel_sheets(file_path)
```

```
## [1] "Biscuits"         "Time Consumption"
## [3] "Weight"
```

The three sheets contain the following information, which need to be combined:

- *Biscuits*: The consumers' evaluation of the 10 products and their assessment on liking, hunger, etc. at different moments of the test.
- *Time Consumption*: The amount of cookies and the time required to evaluate them in each sitting.
- *Weight*: The weight associated to each cookie.

Let's start by combining *Time Consumption* and *Weight* so that we can compute the total weight of biscuits eaten by each consumer in each sitting. In this case, the joining procedure is done by `Product` since the weight is only provided for each product. The total weight eaten (`Amount`) is then computed by multiplying the number of cookies eaten (`Nb biscuits`) by `Weight`:

```
time <- read_xlsx(file_path, sheet = "Time Consumption")
weight <- read_xlsx(file_path, sheet = "Weight")

consumption <- time %>%
    full_join(weight, by = "Product") %>%
    mutate(Amount = `Nb biscuits` * Weight)
```

As can be seen, the `Weight` information stored in the *Weight* sheet has been replicated every time each sample has been evaluated by another respondent.

The next step is then to merge this table to `Biscuits`. In this case, since both data set contain the full evaluation of the cookies (each consumer evaluating each product), the joining procedure needs to be done by `Judge` and `Product` simultaneously. A quick look at the data shows two important things:

1. In *Biscuits*, the consumer names only contain the numbers whereas in `consumption`, they also contain a `J` in front of the name: This needs to be fixed as the names need to be identical to be merged, else they will be considered separately and missing values will be introduced. In practice, this will be done by mutating `Consumer` and by pasting a `J` in front of the number using the function `paste0()`.
2. The names that contain the product (`Samples` and `Product`) and consumers (`Consumer` and `Judge`) information are different in both data sets. We could rename these columns in one data set to match the other, but instead we will keep the two names and inform it

within `full_join()`. This is done through the `by` parameter as follows: `"name in dataset 1" = "name in dataset 2"`

```
biscuits <- read_xlsx(file_path, sheet = "Biscuits") %>%
  mutate(Consumer = str_c("J", Consumer)) %>%
  full_join(consumption,
            by = c("Consumer" = "Judge", "Samples" = "Product"))
```

The three data sets are now flawlessly joined into one that can be further manipulated and/or analyzed.

Since dplyr 1.1.0 release, a new option called 'multiple' has been introduced. This option controls the behaviour of `*_join()` functions when one value from one table matches multiple values from the other table (e.g. our example joining time and weight). If multiple is set to "All" (by default), the values are being repeated for each possible association. Yet, depending on the situation, it is possible to prompt an error, a warning, or to return a table with only one value (whether it is the first/last match, or any matching value) when values should not be duplicated.

# 5

# *Data Visualization*

One of the main goal of data analysis is to produce results. Although informative, such results are only impactful if they can be well-communicated. It is hence of utmost importance to present them in a neat way, often through visuals. In this chapter, two forms of visuals (namely tables and graphs) are being generated using R. Although some design principles are being tackled, the aim of this chapter is to mainly focus on the *how to?* rather than on the design itself. Although R already comes with tools for building and printing tables and graphs, we opt for using additional packages ({`flextable`} and {`gt`} for tables, {`ggplot2`} for graphs) as they provide more flexibility and possibilities.

## 5.1 Introduction

Tables and graphs are the two fundamental vehicles to communicate information clearly and effectively. They are useful visual elements to summarize and organize information to show patterns and relationships. Tables and graphs allow the audience/reader to easily and quickly get a clear idea of the data findings, make comparisons, get insights from it and ultimately, draft conclusions without much effort.

The best medium of communication, whether a table, a bar chart, a line chart, or a radar plot, will highly depend on the type of data, the amount of data to be displayed (e.g. number of attributes or samples), and the purpose of the analysis.

Usually, tables are meant to be read, so they are ideal when you have data that cannot be easily presented by other communication elements or when the data require more specific attention. However, if you encounter a situation where you have a very long and/or wide table, (which is common in sensory

DOI: 10.1201/9781003028611-5

and consumer studies), other vehicle of communication should be considered. The same remark also applies to graphical visualization, and if you have very little data to display, tables might be best suited.

Sometimes (if not often) you have to play with your data, and test displaying it as a table or different types of graphs, before deciding which one suits best. As practical advice, do not hesitate to ask colleagues for feedback as having an external point of view often helps. Remember, to select the best way to communicate your data, you must understand the needs of your audience, the purpose for which various forms of display can be effectively used, but also the strengths and weaknesses of each type of data representation considered.

## 5.2 Design Principles

Regardless of the way you decide to display your data, you must understand visual perception and its application in graphical communication. It is important to spend some time with the design and aesthetic aspects of your visualization. You should be able to recognize smart design by becoming familiar with some aspects and examples of great design. Inattention to visual design such as tables with improper alignment of numbers and excessive use of lines and fill colors can greatly diminish their effectiveness. In other words, when used adequately, design should help you communicate your results by *clarifying* it and not distract your audience from it.

Some important pre-attentive aspects that you should be aware of will be presented in this section, but to read more about visual perception and graphical communication, as well as some examples of great design, we strongly recommend *Storytelling with Data* by Cole Nussbaumer Knaflic (Knaflic, 2015) and *Show me the Number: Designing Table and Graphs to Enlighten* by Stephen Few (Few, 2012).

Since a *picture is worth a thousand words*, let's demonstrate the difference between pre-attentive and attentive processing using an example provided by Stephen Few in his book *Show me the Number: Designing Table and Graphs to Enlighten*. First, take a look at the numbers below and determine, as quickly as you can, how many times the number 5 appears:

98734979027513649782316497802316487415113697412369846321
12346879231648791300023665698774646821397486248797964312
12369874474578962341680021364789613469174312679812439612
12332146987412361789461230502135467980213648126498731203

This appears to be a tedious task and it most likely took you a few minutes because it involved **attentive processing**. The list of numbers did not have any hint (also called pre-attentive attributes) that could help you to easily distinguish the number 5 from the other numbers. Hence, you are forced to perform a sequential search throughout the whole list.

Let's do it again, but now using the list of numbers below:

98734979027513649782316497802316487415113697412369846321
12346879231648791300023665698774646821397486248797964312
12369874474578962341680021364789613469174312679812439612
12332146987412361789461230502135467980213648126498731203

This time the task is much easier and you can count the number of times the number 5 appears much faster. This is because we used the **pre-attentive attribute** of color to distinguish number 5, distinguishing it from the rest. This example shows in an easy way the power of pre-attentive attributes for effective visual communication. As stated by Cole Nussbaumer Knaflic in her book *Storytelling with Data*, when we use pre-attentive attributes strategically, we enable our audience to see what we want them to see before they even know they are seeing it!

The various pre-attentive attributes that can be used to draw your audience's attention quickly and create a visual hierarchy of information include **attributes of form**, such as line length, line width, orientation, shape, size, added marks, and enclosure, **attributes of color**, which would be hue and intensity, and also spatial position and motion. Some of the strategies for a smart design for graphical communication described by Cole Nussbaumer include:

- **Highlight the important stuff** – use tools such as bold, italics, underlining, uppercase text, color, and different sizes to draw your audience's attention to what you want them to focus on.
- **Eliminate distractions** – while some elements should be highlighted, unnecessary or irrelevant items or information should be identified to be cut or de-emphasized to minimize your audience's distraction. Get rid of noncritical data or components, things that wouldn't change the main message, and summarize when details are not needed. When a piece of information is necessary to come along with your visualization but is not really message-impacting, you should de-emphasize it – light gray usually works well for that purpose.

Now the main stage is set, let's focus on how to build nice tables and graphs in R.

## 5.3 Table Making

By default, R allows printing matrices or data frames as tables. However, these tables cannot be customized and are only informative. An example of a table can be shown here:

```
##            Name col 1 Name col 2 Name col 3
## Name row 1          1          3          5
## Name row 2          2          4          6
```

To extend table customization, dedicated packages are required. For illustration, the sensory data (stored in *biscuits_sensory_profile.xlsx*) is used. Before starting, let's first load the usual libraries.

```
library(tidyverse)
library(readxl)
library(here)
```

Let's imagine we want to communicate the sensory profiles of the 11 biscuits in a few attributes including `Shiny`, `Sweet`, `Sour`, `Bitter`, and `Salty`. The first step consists of transforming the data to create such results (see 4.2.3 for more details).

```
file_path <- here("data","biscuits_sensory_profile.xlsx")

mean_sensory <- readxl::read_xlsx(file_path, sheet="Data") %>%
  select(Product, Shiny, Sweet, Sour, Bitter, Salty) %>%
  group_by(Product) %>%
  summarize(across(.cols = where(is.numeric),.fns = mean))
```

### 5.3.1 Introduction to {`flextable`}

Now the correct table has been created, let's represent it in a neater way using our first dedicated package called {`flextable`}. Before going too deep in designing the table, let's simply apply the `flextable()` function to `mean_sensory` (after loading {`flextable`} first):

```
library(flextable)

flex_table <- mean_sensory %>%
  flextable()
```

As you can see, the table is being printed in the *Viewer* section of RStudio. Table 5.1 is already better designed, although it is still quite overcrowded that could benefit from some additional design work to look nicer.

**TABLE 5.1**

Basic flextable

| Product | Shiny | Sweet | Sour | Bitter | Salty |
|---|---|---|---|---|---|
| P01 | 41.933333 | 22.20000 | 0.0000000 | 8.000000 | 5.100000 |
| P02 | 39.133333 | 15.80000 | 0.0000000 | 4.933333 | 2.933333 |
| P03 | 30.533333 | 10.40000 | 0.0000000 | 7.800000 | 4.666667 |
| P04 | 42.600000 | 16.60000 | 0.0000000 | 4.266667 | 3.600000 |
| P05 | 13.933333 | 21.00000 | 3.0000000 | 6.733333 | 5.866667 |
| P06 | 6.866667 | 14.73333 | 4.0000000 | 6.533333 | 8.200000 |
| P07 | 32.466667 | 19.46667 | 0.8666667 | 5.000000 | 2.800000 |
| P08 | 8.933333 | 14.26667 | 0.9333333 | 11.066667 | 2.466667 |
| P09 | 3.533333 | 13.33333 | 0.0000000 | 9.266667 | 4.000000 |
| P10 | 26.133333 | 30.46667 | 5.0666667 | 9.666667 | 9.000000 |
| POpt | 16.866667 | 17.73333 | 2.2000000 | 15.866667 | 6.666667 |

A first simple improvement consists of reducing the number of decimals (`colformat_double()`), changing the font size (`fontsize()`, not visible in our output) and type (`bold()` or `italic()`), and aligning the text (`align()`) for instance (Table 5.2).

```
flex_table_design <- flex_table %>%
  colformat_double(digits = 2) %>%
  fontsize(size = 10, part = "all") %>%
  bold(bold = TRUE, part = "header") %>%
  italic(j = -1, italic = TRUE, part = "body") %>%
  align(align = "center", part = "all")
```

**TABLE 5.2**

Flextable – aesthetics improvements

| **Product** | **Shiny** | **Sweet** | **Sour** | **Bitter** | **Salty** |
|---|---|---|---|---|---|
| P01 | *41.93* | *22.20* | *0.00* | *8.00* | *5.10* |
| P02 | *39.13* | *15.80* | *0.00* | *4.93* | *2.93* |
| P03 | *30.53* | *10.40* | *0.00* | *7.80* | *4.67* |
| P04 | *42.60* | *16.60* | *0.00* | *4.27* | *3.60* |
| P05 | *13.93* | *21.00* | *3.00* | *6.73* | *5.87* |
| P06 | *6.87* | *14.73* | *4.00* | *6.53* | *8.20* |
| P07 | *32.47* | *19.47* | *0.87* | *5.00* | *2.80* |
| P08 | *8.93* | *14.27* | *0.93* | *11.07* | *2.47* |
| P09 | *3.53* | *13.33* | *0.00* | *9.27* | *4.00* |
| P10 | *26.13* | *30.47* | *5.07* | *9.67* | *9.00* |
| POpt | *16.87* | *17.73* | *2.20* | *15.87* | *6.67* |

As can be seen, the function names are very intuitive, and so are the options. In particular, it is interesting to see that most functions allow applying changes to the entire table (`part = "all"`), the header only (`part = "header"`), or the body only (`part = "body"`). And even within a part, it is possible to make a sub-selection by selecting the rows (`i`) or columns (`j`) to include or exclude. Here, all the text in the body is set in italic except for the product names, hence the option `j=-1` (read exclude the first column).

After presenting some of the basic aesthetic options, let's go one step further and play around with coloring. For instance, let's imagine we would change the header background and text color, and would want to call the audience's attention by highlighting the optimized formulation. We can also change the colour of the outer lines of the table using `fp_border()` from {`officer`} (Table 5.3). The following code could be used to do this (results are not being saved):

```
flex_table_design %>%
  bg(bg = "black", part = "header") %>%
  color(color = "white", part = "header") %>%
  fontsize(size = 13, part = "header", i = 1) %>%
  color(i = 11, color = "orange", part = "body") %>%
  color(i = 1:10, color = "grey70", part = "body") %>%
  add_header_lines(values = "Sensory Profile of 11 biscuits") %>%
  border_remove() %>%
  border_outer(border=officer::fp_border(color="darkorange", width=2)) %>%
  fix_border_issues()
```

**TABLE 5.3**

Flextable – coloring function

| Sensory Profile of 11 biscuits | | | | | |
|---|---|---|---|---|---|
| Product | Shiny | Sweet | Sour | Bitter | Salty |
| P01 | *41.93* | *22.20* | *0.00* | *8.00* | *5.10* |
| P02 | *39.13* | *15.80* | *0.00* | *4.93* | *2.93* |
| P03 | *30.53* | *10.40* | *0.00* | *7.80* | *4.67* |
| P04 | *42.60* | *16.60* | *0.00* | *4.27* | *3.60* |
| P05 | *13.93* | *21.00* | *3.00* | *6.73* | *5.87* |
| P06 | *6.87* | *14.73* | *4.00* | *6.53* | *8.20* |
| P07 | *32.47* | *19.47* | *0.87* | *5.00* | *2.80* |
| P08 | *8.93* | *14.27* | *0.93* | *11.07* | *2.47* |
| P09 | *3.53* | *13.33* | *0.00* | *9.27* | *4.00* |
| P10 | *26.13* | *30.47* | *5.07* | *9.67* | *9.00* |
| POpt | *16.87* | *17.73* | *2.20* | *15.87* | *6.67* |

Alternatively, we could decide to be more sober by applying other pre-attentive attributes (Table 5.4). For instance, the size of the table can be adjusted and an horizontal line can be added to delimit the optimal sample from the other. For the latter part, customization of the line can be made using the function `fp_border()` from the {officer} package.[1]

```
library(officer)

flex_table_design %>%
  hline(i=10, border=fp_border(color="grey70", style="dashed")) %>%
  autofit()
```

**TABLE 5.4**

Flextable – line customization

| Product | Shiny | Sweet | Sour | Bitter | Salty |
|---|---|---|---|---|---|
| P01 | 41.93 | 22.20 | 0.00 | 8.00 | 5.10 |
| P02 | 39.13 | 15.80 | 0.00 | 4.93 | 2.93 |
| P03 | 30.53 | 10.40 | 0.00 | 7.80 | 4.67 |
| P04 | 42.60 | 16.60 | 0.00 | 4.27 | 3.60 |
| P05 | 13.93 | 21.00 | 3.00 | 6.73 | 5.87 |
| P06 | 6.87 | 14.73 | 4.00 | 6.53 | 8.20 |
| P07 | 32.47 | 19.47 | 0.87 | 5.00 | 2.80 |
| P08 | 8.93 | 14.27 | 0.93 | 11.07 | 2.47 |
| P09 | 3.53 | 13.33 | 0.00 | 9.27 | 4.00 |
| P10 | 26.13 | 30.47 | 5.07 | 9.67 | 9.00 |
| POpt | 16.87 | 17.73 | 2.20 | 15.87 | 6.67 |

In some situations, applying some design options might mess up the appearance of your table, in particular its border lines. If that should happen to you, just apply the function `fix_border_issues()` at the end of your code to solve it.

Lastly, you can use conditional formatting if you want to highlight some specific values (e.g. values for `Sweet` that are below 20 should be colored in blue, and above 30 in red), as in the example below (Table 5.5):

```
color_code <- ifelse(mean_sensory$Sweet <= 20, "blue",
                     ifelse(mean_sensory$Sweet >= 30, "red", "black"))

flex_table_design %>%
  color(j=~Sweet, color=color_code)
```

[1] More information regarding the {officer} package are provided in Section 6.2.2.

**TABLE 5.5**
Flextable – conditional formatting

| Product | Shiny | Sweet | Sour | Bitter | Salty |
|---|---|---|---|---|---|
| P01 | 41.93 | 22.20 | 0.00 | 8.00 | 5.10 |
| P02 | 39.13 | 15.80 | 0.00 | 4.93 | 2.93 |
| P03 | 30.53 | 10.40 | 0.00 | 7.80 | 4.67 |
| P04 | 42.60 | 16.60 | 0.00 | 4.27 | 3.60 |
| P05 | 13.93 | 21.00 | 3.00 | 6.73 | 5.87 |
| P06 | 6.87 | 14.73 | 4.00 | 6.53 | 8.20 |
| P07 | 32.47 | 19.47 | 0.87 | 5.00 | 2.80 |
| P08 | 8.93 | 14.27 | 0.93 | 11.07 | 2.47 |
| P09 | 3.53 | 13.33 | 0.00 | 9.27 | 4.00 |
| P10 | 26.13 | 30.47 | 5.07 | 9.67 | 9.00 |
| POpt | 16.87 | 17.73 | 2.20 | 15.87 | 6.67 |

Other illustrations of the use of {flextable} are provided in Section 6.2.2. For curious readers who want to get a deeper look into all the possibilities provided by this package, we refer them to the book *Using the flextable R package*[2] by David Gohel and to *Flextable gallery*[3] for inspiration.

### 5.3.2 Introdution to {gt}

As an alternative to {flextable}, the {gt} package can also be considered as it also produces nice-looking tables for reports or presentations. Let's first install (if needed) and load the {gt} package.

```
library(gt)
```

Focusing now on the consumer study (*biscuits_consumer_test.xlsx*), let's display a table with the average number of biscuits (for each variant) consumers ate and their corresponding eating time. To do so, we first need to transform the time columns (expressed as min and s) to a double format and express them in minutes. Then, we can group them by product to get the average for the time spent and the number of biscuits eaten.

```
file_path <- here("data","biscuits_consumer_test.xlsx")
```

[2] https://ardata-fr.github.io/flextable-book/index.html
[3] https://ardata.fr/en/flextable-gallery/

```
mean_consumer <- readxl::read_xlsx(file_path,
                                   sheet="Time Consumption") %>%
  dplyr::select(Product, `Time (min)`, `Nb biscuits`) %>%
  separate(`Time (min)`, c("Min", "Sec"), sep="min") %>%
  mutate(across(c("Min","Sec"), as.numeric)) %>%
  mutate(Time = Min+Sec/60) %>%
  group_by(Product) %>%
  summarise(across(c("Time", "Nb biscuits"), mean, na.rm = TRUE)) %>%
  ungroup()
```

Now that the data are ready, we can display it with some basic adjustments to make it look nicer (Table 5.6).

**TABLE 5.6**
gt Table

**Cons. time and nb. of biscuits eaten**
*Average taken from 99 consumers*

| Product | Time | Nb biscuits |
|---|---|---|
| 1 | 6.47 | 3.94 |
| 2 | 6.74 | 4.00 |
| 3 | 6.46 | 2.91 |
| 4 | 6.48 | 2.61 |
| 5 | 6.57 | 4.06 |
| 6 | 6.45 | 3.42 |
| 7 | 6.55 | 3.58 |
| 8 | 6.23 | 2.77 |
| 9 | 6.24 | 3.01 |
| 10 | 6.95 | 4.41 |

```
mean_consumer %>%
  gt () %>%
  cols_align(align = "center", columns = everything()) %>%
  fmt_number(columns = c("Time", "Nb biscuits") , decimals = 2) %>%
  tab_header(title = md("**Cons. time and nb. of biscuits eaten**"),
             subtitle = md ("*Average taken from 99 consumers*"))
```

Note that we used *Markdown* to style the title and subtitle by wrapping the values passed to the title or subtitle with the `md()` function. In Markdown, `**text**` writes the text in bold and `*text*` in italic.

The {`gt`} package offers several resources to make beautiful tables. Let's illustrate this by focusing on the average number of biscuits eaten only since the average consumption time is very similar across products. The idea is to use pre-attentive attributes for the audience to clearly see which samples were the most popular (i.e. the most eaten) and which one were not. Let's first prepare the data and calculate the overall time consumption considering all products.

```
mean_consumer_2 <- mean_consumer %>%
  dplyr::select(- "Time") %>%
  arrange(desc (`Nb biscuits`))
```

Now that the data are ready, we can display using a similar style as before in which we add some color-code to accentuate products' consumption. We will also add a `note` to the table that expresses the average time used to consume the biscuits. So let's start with creating the table:

```
note <- str_c("Avg. consumption time: ",
              round(mean(mean_consumer$Time),2), " min")

consumption <- mean_consumer_2 %>%
  gt () %>%
  cols_align(align = "center", columns = everything()) %>%
  fmt_number(columns = "Nb biscuits" , decimals = 2) %>%
  tab_header(title = md ("**Number of biscuits eaten**"),
             subtitle = md ("*Average taken from 99 consumers*")) %>%
  tab_source_note(source_note = note)
```

Now, let's color code the cells based on the average number of biscuits eaten. To color code, the range of average number of biscuits eaten is required. Then, we can use the `col_numeric()` function from the {`scales`} package to generate the colors of interest (in practice, we provide the color for the minimum and maximum, and the function generates automatically all the colors in between to create the gradient).

```
library(scales)

nb_range <- range(mean_consumer_2$`Nb biscuits`)

consumption %>%
    data_color(columns=`Nb biscuits`,
               colors=col_numeric(c("#FEF0D9","#990000"),
                                  domain=nb_range, alpha=0.75))
```

Applying this strategy of coloring the number of biscuits eaten according to their range makes the table nicer and easier to get insights from (Table 5.7). In our case, we can quickly see the groups of products based on their average consumption: Product 10 is the most eaten, followed by a group that includes products 5, 2, and 1. At last, samples 8 and 4 are the least-consumed samples.

**TABLE 5.7**
gt Table – coloring function

| **Number of biscuits eaten** | |
|---|---|
| *Average taken from 99 consumers* | |
| Product | Nb biscuits |
| 10 | 4.41 |
| 5 | 4.06 |
| 2 | 4.00 |
| 1 | 3.94 |
| 7 | 3.58 |
| 6 | 3.42 |
| 9 | 3.01 |
| 3 | 2.91 |
| 8 | 2.77 |
| 4 | 2.61 |
| Avg. consumption time: 6.52 min | |

Although the package {`gt`} proposed some nice features, additional options are provided by its extension package called {`gtExtras`} which provides additional themes, formatting capabilities, opinionated diverging color palette, extra tools for highlighting values, possibility of embedding bar plots in the table, etc. For more information, please check gtExtras.[4]

To illustrate one of the possible use of {`gtExtras`}, let's twist the previous table as following: Since each consumer was provided with a maximum of 10 biscuits, let's transform the average consumption into percentages. We can then recreate the previous table in which we also add a bar chart based on the percentages (Table 5.8):

[4] https://jthomasmock.github.io/gtExtras/

```
library(gtExtras)

mean_consumer_2 %>%
  mutate(`Nb biscuits (%)` = 100*(`Nb biscuits`/10)) %>%
  gt () %>%
  cols_align(align = "center", columns = everything()) %>%
  fmt_number(columns = "Nb biscuits" , decimals=2) %>%
  tab_header(title = md ("**Number of biscuits eaten**"),
             subtitle = md ("*Average taken from 99 consumers*")) %>%
  tab_source_note(source_note = note) %>%
  gt_plt_bar_pct(`Nb biscuits (%)`, scaled=TRUE)
```

**TABLE 5.8**
gtExtra Table

**Number of biscuits eaten**

*Average taken from 99 consumers*

| Product | Nb biscuits | Nb biscuits (%) |
|---|---|---|
| 10 | 4.41 | |
| 5 | 4.06 | |
| 2 | 4.00 | |
| 1 | 3.94 | |
| 7 | 3.58 | |
| 6 | 3.42 | |
| 9 | 3.01 | |
| 3 | 2.91 | |
| 8 | 2.77 | |
| 4 | 2.61 | |

Avg. consumption time: 6.52 min

In this section, due to their simplicity and flexibility, we emphasized the use of {flextable} and {gt} to build beautiful tables for your reports. However, there are other alternatives including {kable} and {kableExtra}, or {huxtable} for readers that are not yet fully satisfied.

## 5.4 Chart Making

"A picture is worth 1000 words". This saying definitely applies to Statistics as well, since visual representation of data often appears clearer than the values themselves stored in a table. It is hence no surprise that R is also a powerful tool for building graphics.

In practice, there are various ways to build graphics in R. In fact, R itself comes with an engine for building graphs through the `plot()` function. An extensive description can be found in *R Graphics* by Murrell (2011). Due to its philosophy, its simplicity, and the point of view adopted in this book, we will limit ourselves to graphics built using the {`ggplot2`} package.

### 5.4.1 Philosophy of {ggplot2}

{`ggplot2`} belongs to the {`tidyverse`} and was developed by H. Wickham and colleagues at RStudio. It is hence no surprise that a lot of the procedures that we are learning throughout this book also applies to {`ggplot2`}. More generally, building graphics with {`ggplot2`} fits very well within the pipes (`%>%`) system from {`magrittr`}. In fact, {`ggplot2`} works with its own piping system that uses the `+` symbol instead of `%>%`.

In practice, {`ggplot2`} is a multi-layer graphical tool, and graphics are built by adding layers to existing graphs. The advantage of such procedure is that `ggplot` objects are not fixed: They can be printed at any time, and can still be improved by adding other layers if needed. To read more about {`gglot2`} and its philosophy, please refer to *A Layered Grammar of Graphics*[5] by Hadley Wickham.

Note that since building graphics is limited to one's imagination, it is not possible to tackle each and every possibility offered by {`ggplot2`} (and its extensions). For that reason, we prefer to focus on how {`ggplot2`}works, and by using as illustration throughout the book examples of graphics that are useful in Sensory and Consumer Science. This should be more than sufficient to get you started, and should cover 90% of your daily needs. Still, if that should not be sufficient, we invite you to look into the online documentation or to references such as *The R Graph Gallery.*[6]

### 5.4.2 Getting Started with {ggplot2}

To use {`ggplot2`}, it needs to be loaded. This can either be done directly using:

```
library(ggplot2)
```

[5] http://vita.had.co.nz/papers/layered-grammar.pdf
[6] https://r-graph-gallery.com/

However, as said before, this step is not always needed since it is part of {tidyverse}: At the start of this chapter, when we loaded the {tidyverse} package, we also loaded {ggplot2}.

To illustrate the use of {ggplot2}, both the sensory data (*biscuits_sensory_profile.xlsx*) and the number of biscuits eaten by each respondents (*biscuits_consumer_test.xlsx*) are used. Although these files have already been loaded in Section 5.3, let's reload them:

```
file_path <- here("data","biscuits_sensory_profile.xlsx")

p_info <- readxl::read_xlsx(file_path, sheet="Product Info") %>%
  dplyr::select(-Type)

sensory <- readxl::read_xlsx(file_path, sheet="Data") %>%
  inner_join(p_info, by="Product") %>%
  relocate(Protein:Fiber, .after=Product)

file_path <- here("Data","biscuits_consumer_test.xlsx")

Nbiscuits <- readxl::read_xlsx(file_path, sheet="Time Consumption") %>%
  mutate(Product = str_c("P", Product)) %>%
  rename(N = `Nb biscuits`)
```

To initiate a graph, the function `ggplot()` is called. Since the data to be used are stored in `sensory`, `ggplot()` is applied on `sensory`:

```
p <- ggplot(sensory)
```

Running this line of code generates an empty graphic stored in `p`. This is logical since no layers have been added yet. So let's imagine we want to look at the overall relationship between `Sticky` and `Melting`. To evaluate this relationship, a scatter plot with `Sticky` in the X-axis and `Melting` in the Y-axis is created (Figure 5.1). To do so, two types of information are required:

- the type of visual (here a scatter point);
- the information regarding the data to plot (what should be represented).

Such information can be provided as such:

```
p + geom_point(aes(x=Sticky, y=Melting))
```

This code adds a layer that consists of points (defined by `geom_point()`) in which the X-axis coordinates are defined by `Sticky` and the Y-axis coordinates by `Melting`, as defined through aesthetics (or `aes()`). This layer is added to the already existing graph `p`.

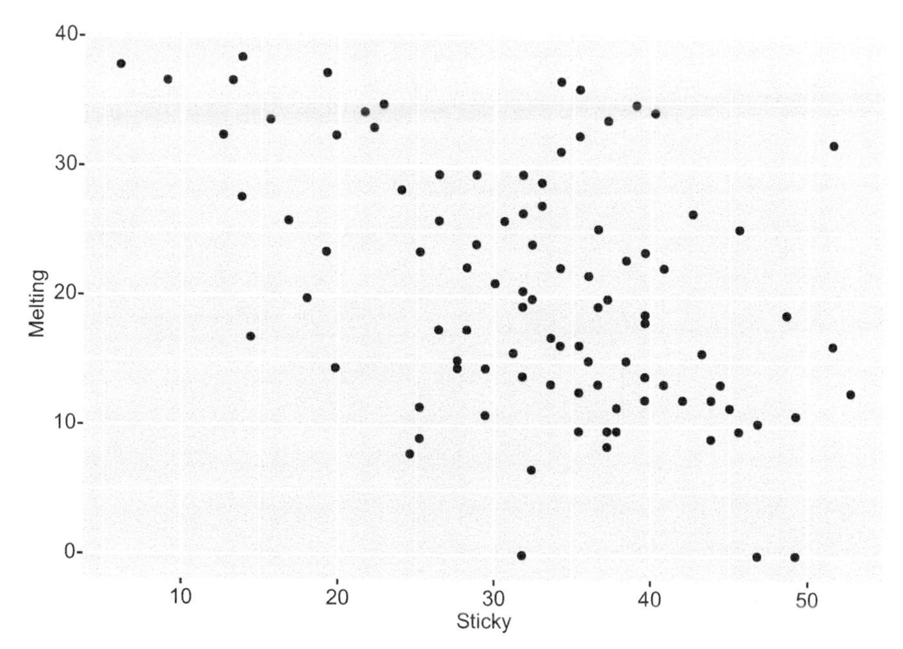


**FIGURE 5.1**
Scatter plot.

### *Introduction to Aesthetics*

In the previous example, one can notice that many points are being printed. This surprising result is logical since `sensory` contains the raw sensory data, meaning that there are as many points as there are assessors evaluating products.

Let's color the points per products to see if we can see any patterns. Since the color code is specific to the data (more precisely to `Product`), it should be informed within the aesthetics by adding `colour=Product` within `aes()` (Figure 5.2):

```
p + geom_point(aes(x=Sticky, y=Melting, colour=Product))
```

As you can see, any parameters provided within `aes()` may depend on a variable (e.g. `colour` in the previous example). If for any reasons, a specific setting should uniformly be applied to all the elements of the graph, then it should be defined outside `aes()`.

Let's illustrate this by providing a simple example in which we change the type of the dots from circle to square using `pch` and increase their size using `cex` (Figure 5.3):

```
p + geom_point(aes(x=Sticky, y=Melting, colour=Product), pch=15, cex=5)
```

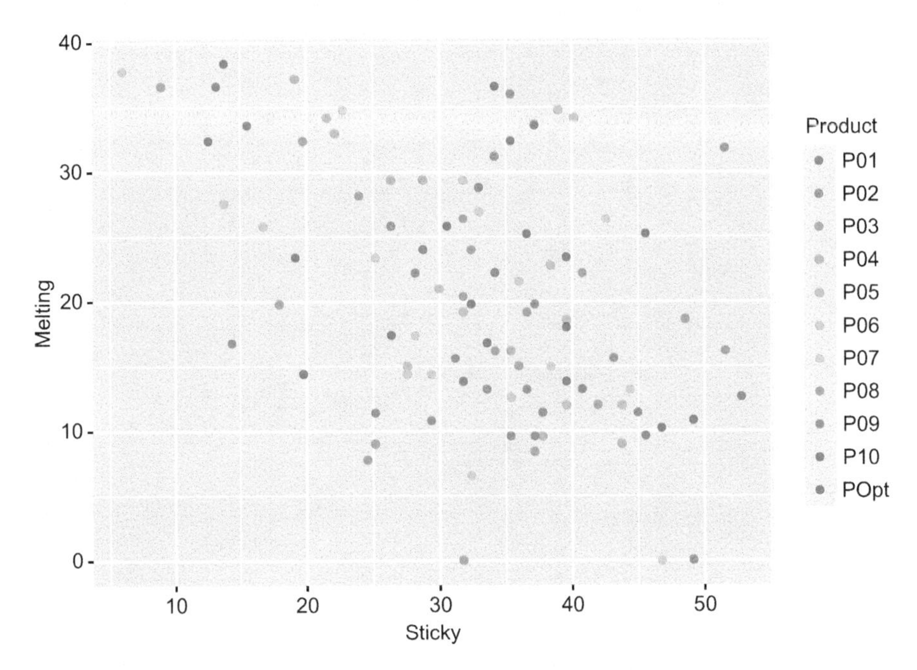


**FIGURE 5.2**
Scatter plot with emphasis on the color through `aes()`.

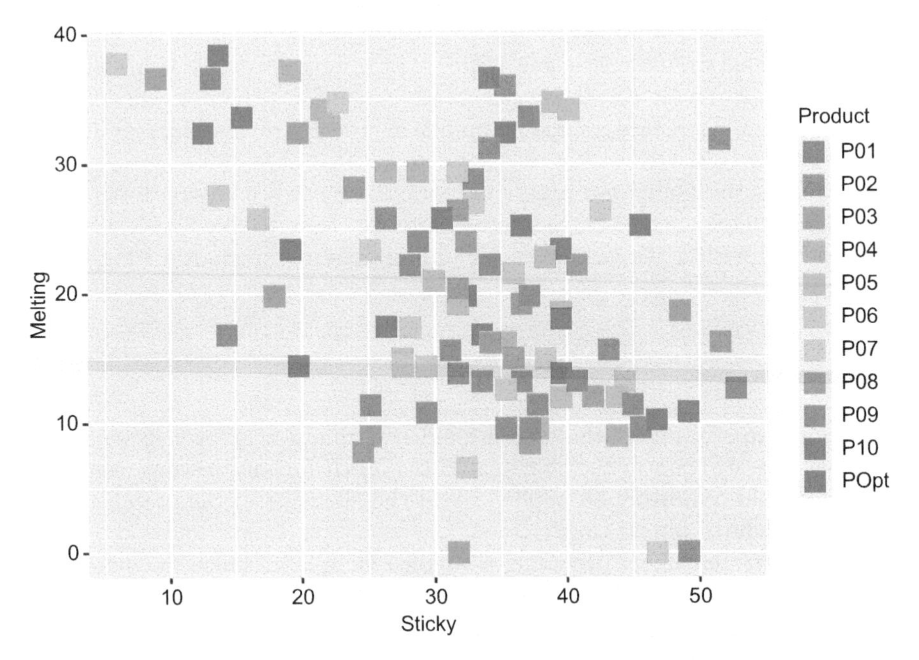


**FIGURE 5.3**
Scatter plot with emphasis on the type of dots through `pch` and `cex`.

Regardless of the products, all the points are now shown as large squares.

Depending on the geom_*() function considered, different parameters should be informed within aes(). Here is a list of the most common aes() you would use:

- x, y, z, provides the coordinates on the X, Y, Z dimensions, respectively;
- colour/color, fill controls for the color code[7] that is being applied to the different elements of a graph;
- group makes the distinction between points that belong to different groups[8];
- text, label prints text/labels on the graph;
- size controls the size of the element (this should preferably be used with numerical variables).

It may not be clear yet on how those aesthetics work, but don't worry, many examples illustrating the use of these various types of aesthetics are provided throughout the book.

### *Introduction to* geom_*() *Functions*

Since {ggplot2} is a multi-layer graph, let's add another layer. For example, the name/code of the panelists associated to each point can be printed.

Such procedure is done through the use of another geom_*() function in geom_text()[9] which requires in aes() the position of the labels (x and y) as well as the label itself.

To avoid having the label overlapping with the point, the text is slightly shifted vertically using nudge_y. For simplicity, let's rebuild the graph from the start (Figure 5.4):

```
ggplot(sensory)+
  geom_point(aes(x=Sticky, y=Melting, colour=Product))+
  geom_text(aes(x=Sticky, y=Melting, label=Judge), nudge_y=1)
```

One interesting remark is that some information required in aes() is being repeated across the different geom_*() used. Such writing can be simplified by

---

[7] You can also use alpha to control for the transparency of the elements by defining values between 0 (completely transparent) to 1 (no transparency).

[8] Note that colour and fill are specific cases of groups as they additionally provide a visual cue on the groups through the color code.

[9] Try using geom_label() instead of geom_text() to see the difference between these two.

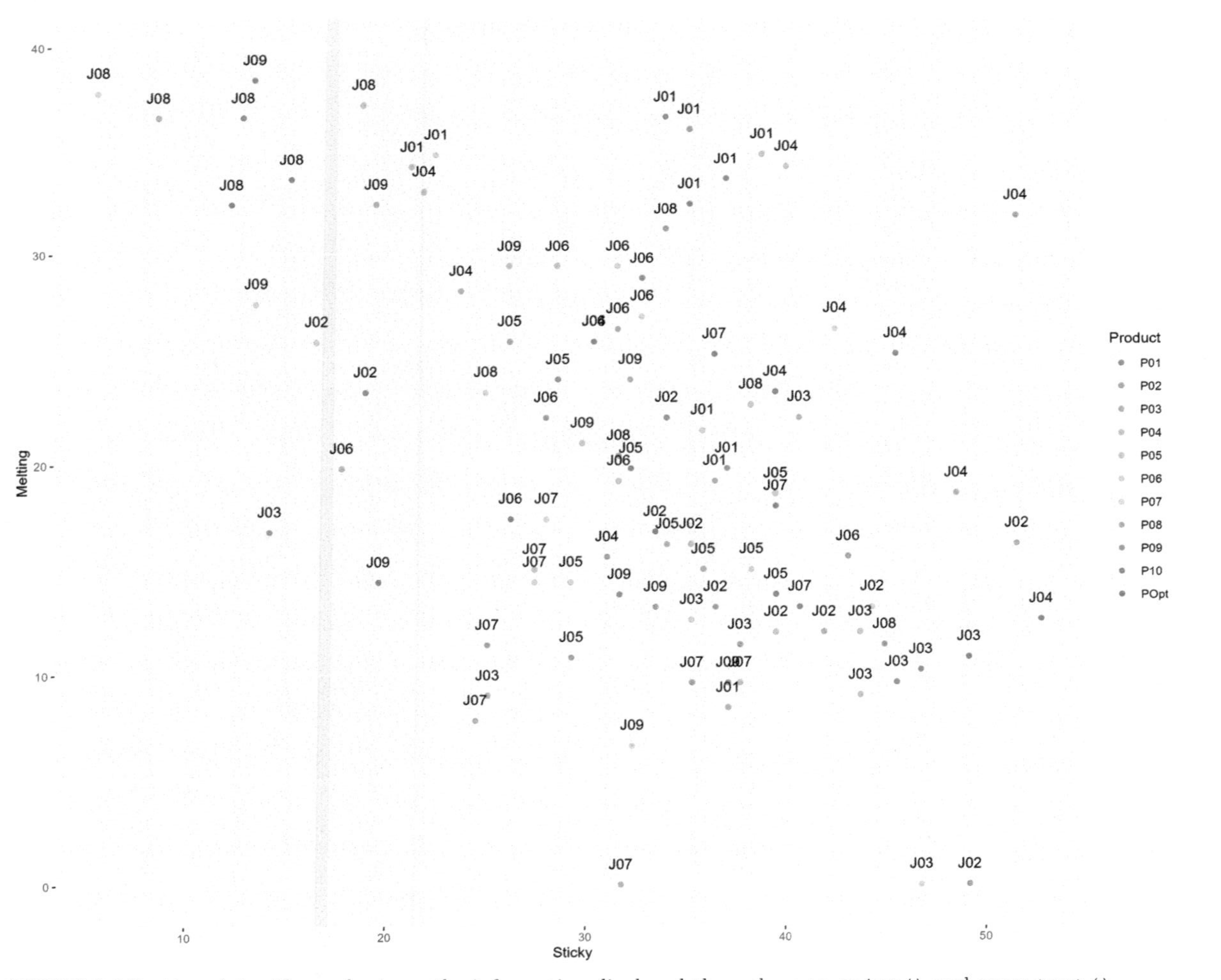


**FIGURE 5.4** Scatter plot with emphasis on the information displayed through `geom_point()` and `geom_text()`.

providing the `aes()` information that applies to all `geom_*()` to the original `ggplot()` call.[10] The previous code hence can be simplified to:

```
p <- ggplot(sensory, aes(x=Sticky, y=Melting, label=Judge))+
  geom_point(aes(colour=Product))+
  geom_text(nudge_y=1)
```

With this new code, you'll notice that:

- `x` and `y` are automatically applied to both `geom_point()` and `geom_text()`;
- although `label` is only relevant for `geom_text()`, it can still be provided at the beginning as it will be ignored by `geom_point()`;
- `colour` should only be provided within `geom_point()` else the text would also be colored according to `Product` (which we do not want here);
- `nudge_y` is defined outside `aes()` as it applies to all the text.

Since the graphics look at the relationship between two quantitative variables, let's add another layer to the previous graph that shows the regression line (Figure 5.5):

```
line_p <- p +
  geom_smooth(method=lm, formula="y~x", se=FALSE)
```

This code adds a regression line to the graphic. It is built using the `lm()` engine in which the simple linear regression model `y~x` is fitted. This result is somewhat surprising since we have not run any regression yet, meaning that `geom_smooth()` is performing this analysis in the background by itself.

In practice, most `geom_*()` function comes with a statistical procedure attached to it. This means that on the raw data, the `geom_*()` function calls its `stat_*()` function that runs the corresponding analysis. In the previous example, `geom_smooth()` calls `stat_smooth()`.

Let's illustrate this concept again using another example: Bar chart that is applied on the data stored in `Nbiscuits`. Here, we want to see the distribution (through bar charts) of the number of biscuits eaten per consumer. A quick look at the data shows that some respondents ate portions of the cookies. To

[10] Intrinsically, this is what is done with `sensory` which is only mentioned within `ggplot()` and is not repeated across the different `geom_*()` functions.

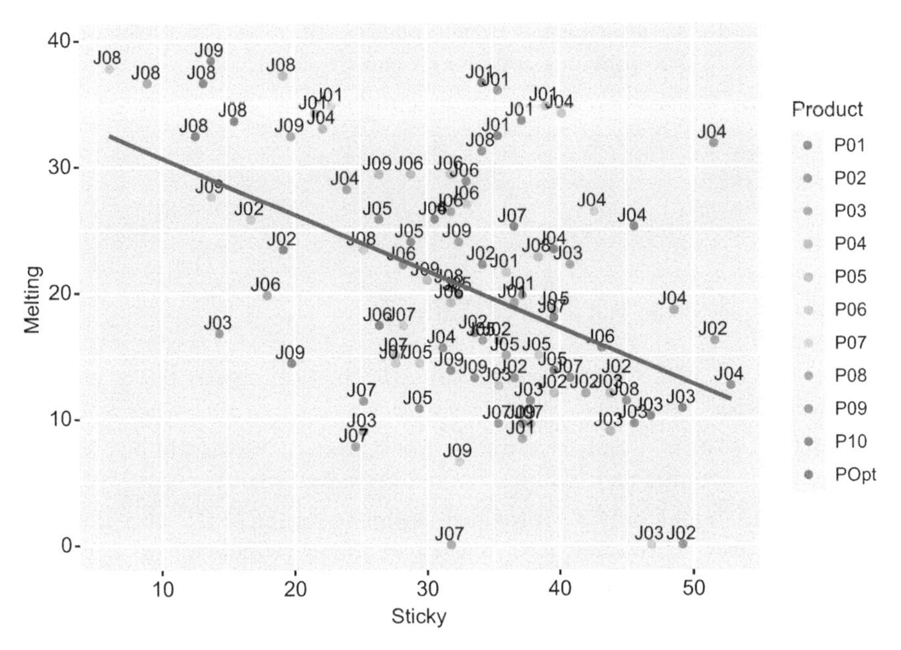


**FIGURE 5.5**
Scatter plot with regression line through `geom_smooth()`.

simplify the analysis, let's consider the total number of entire cookies eaten: If a respondent has eaten say 3.5 biscuits, it will be rounded down to 3 full cookies.

```
Nbiscuits <- Nbiscuits %>%
  mutate(N = floor(N))
```

To create such distribution, a first solution consists of counting for each product how many respondents ate 0 biscuit, 1 biscuit, 2 biscuits, etc. This is automatically done using `geom_bar` and `stat="count"`. The parameter `position="dodge"` is used to get the results per biscuit side by side rather than stacked up vertically (value by default) (Figure 5.6):

```
bar_p <- ggplot(Nbiscuits, aes(x=N, fill=Product)) +
  geom_bar(stat="count", position="dodge")
```

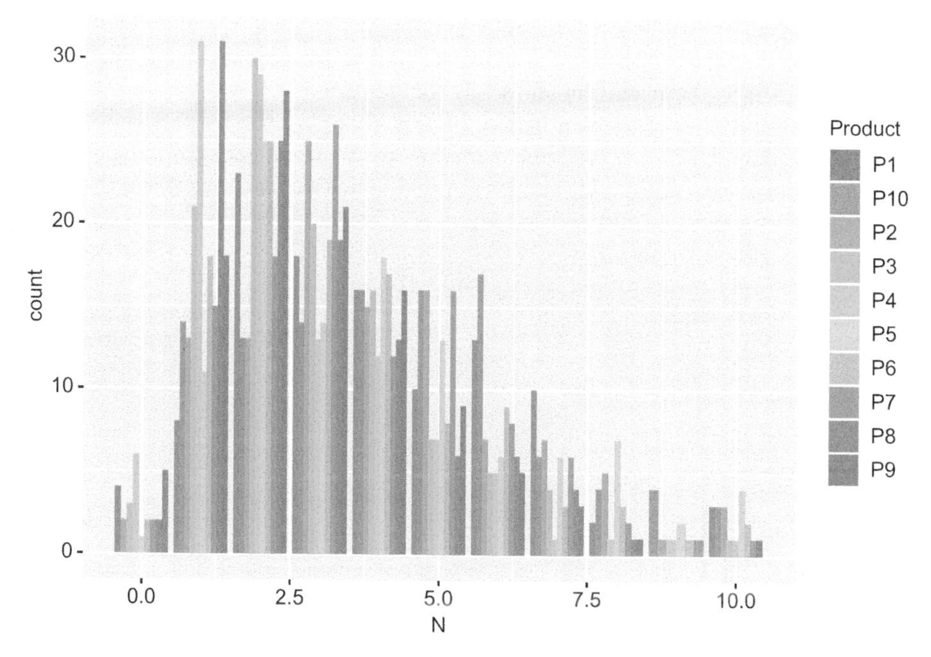


**FIGURE 5.6**
Bar chart displaying the distribution using `geom_bar()`.

In the background, this corresponds to grouping the data by `Product`, summarizing the results by counting `N`, and then performing `geom_bar()` in which no transformation is required (we set `stat="identity"`)[11]:

```
Nbiscuits %>%
  count(Product, N) %>%
  ggplot(aes(x=N, y=n, fill=Product))+
  geom_bar(stat="identity", position="dodge")
```

As can be seen, these two graphics are identical.

## Making Graphs Pretty

In the two previous graphs generated (stored in `line_p` and `bar_p`), some features can be changed to produce clearer visualizations. Currently, the background is gray with vertical and horizontal white lines, the legend is

[11] This code could even be simplified by using `geom_col()` which corresponds to `geom_bar()` with `stat="identity"` as default.

positioned on the right side, the axis is defined based on the data itself (and so are the axis titles), there is no title, etc. All these points (and many more) can be modified, as it will be shown in this section.

Let's start with a quick win by completely changing the overall appearance of the graphic. To do so, predefined *themes* with preset backgrounds (with or without lines, axis lines, etc.) can be applied. The two themes we use the most are `theme_minimal()` and `theme_bw()` (see *Complete themes*[12] for a complete list of predefined themes.)

Let's start with improving `bar_p` using `theme_minimal()` (Figure 5.7):

```
bar_p <- bar_p+
  theme_minimal()
```

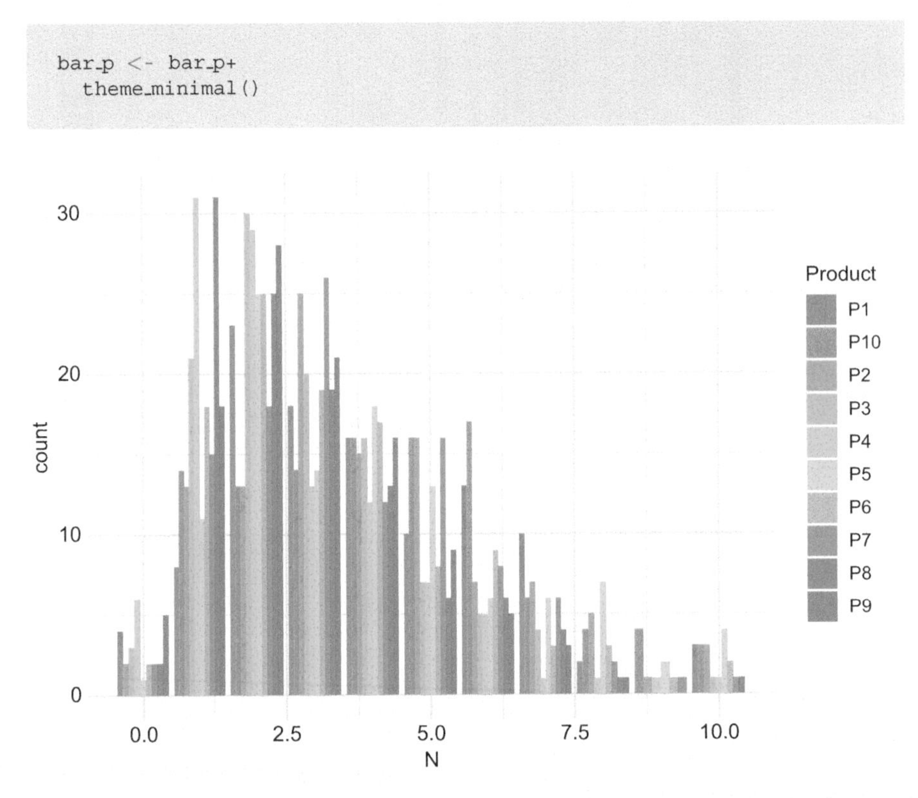


**FIGURE 5.7**
Bar chart with emphasis on the display through `theme()`.

Rather than using predefined themes (or to complement predefined themes), the different parameters of the graph can be controlled through `theme()`.

Let's modify the axes by changing their names and by applying more logical breaks. For instance, the limits of the x-axis can be extended to $-1$ and 11 to ensure that all the histograms are visible, else R removes some and returns a warning: *Removed 10 rows containing missing values* (Figure 5.8).

[12] hhttps://ggplot2.tidyverse.org/reference/ggtheme.html

```
bar_p <- bar_p +
  scale_x_continuous(name="Number of Biscuits eaten",
                     breaks=seq(0,10,1),
                     labels=c("None", 1:9, "All of them"),
                     limits=c(-1,11))+
  ylab("Number of Respondents")
```

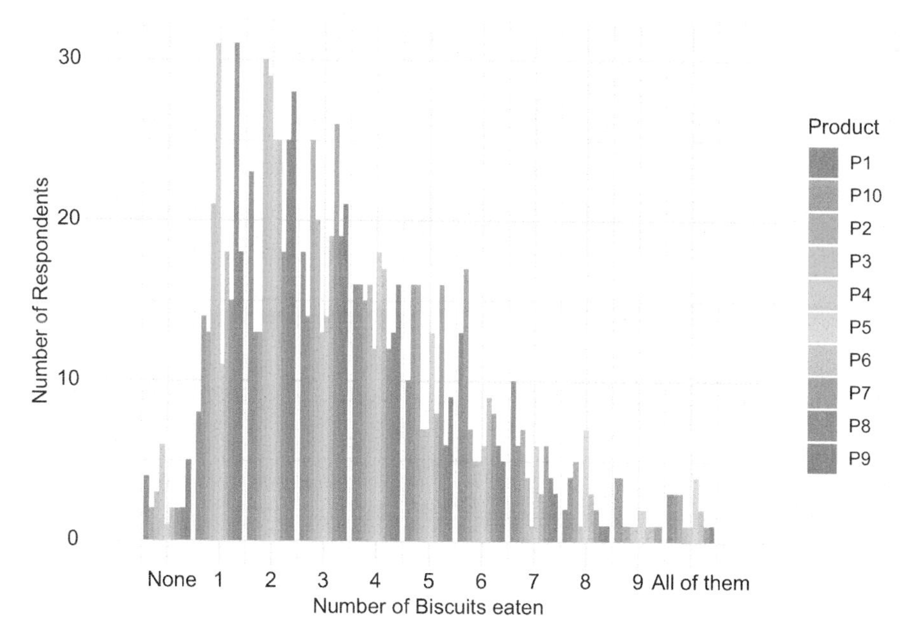


**FIGURE 5.8**
Bar chart with emphasis on the display through `scale_x_continuous()` and `ylab()`.

Last but not least, a title is being added to the graph using `ggtitle()` (Figure 5.9):

```
bar_p <- bar_p +
  ggtitle("Distribution of the number of biscuits eaten",
          "(Results are split per biscuit type)")
```

Let's apply a similar transformation to `line_p`. Here, we are aiming a more *realistic* plot using Cartesian coordinates, a nice theme, no legend, and a title to the graph (Figure 5.10).

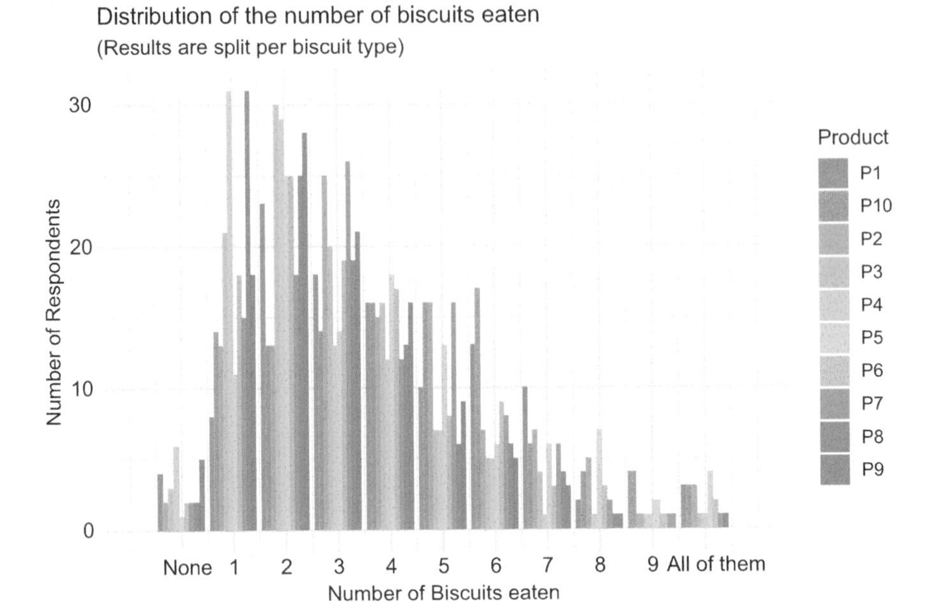


**FIGURE 5.9**
Bar chart with emphasis on the display through `ggtitle()`.

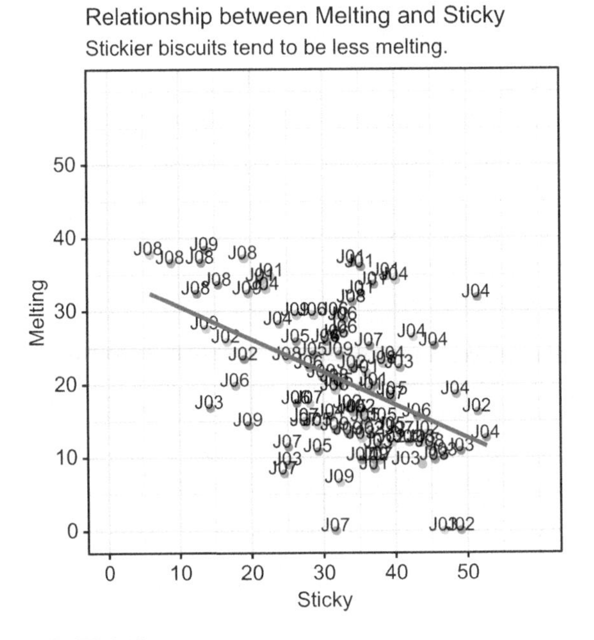


**FIGURE 5.10**
Scatter plot with emphasis on the display through `theme()`, `scale_*_continuous()`, `coord_fixed()`, `ggtitle()` and `guides()`.

```
line_p <- line_p +
  theme_bw()+
  scale_x_continuous(breaks=seq(0,50,10), limits=c(0,60))+
  scale_y_continuous(breaks=seq(0,50,10), limits=c(0,60))+
  coord_fixed()+
  ggtitle("Relationship between Melting and Sticky",
          "Stickier biscuits tend to be less melting.")+
  guides(colour="none")
```

### 5.4.3 Common Charts

You have now an overview of the basics of {ggplot2} and its philosophy. You'll find plenty of other examples throughout this book to help you develop your skills in building graphics in R.

Since making an exhaustive list of plots that are relevant in sensory and consumer science is out of the scope of this book, it is not going to be further developed here. Yet, here is a summary of the `geom_*()` that are commonly used:

- Scatter points:
  - `geom_point()`: create a scatter point (see example Section 5.4.2)
- Line charts:
  - `geom_line()`: create a line that connects points;
  - `geom_smooth()`: add a regression line (see example Section 5.4.2);
  - `geom_hline()` (resp. `geom_vline()`): add a horizontal (resp. vertical) line using `yintercept` (resp. `xintercept`);
  - `geom_segment()`: draw a segment going from (x;y) to (xend;yend).[13]
- Bar charts:
  - `geom_col()` and `geom_bar()`: produce bar charts by either using the raw values or by computing the frequencies first (see example Section 5.4.2);
  - `geom_histogram()` and `geom_freqpoly()`: work in a similar way as `geom_bar()` except that it divides the x axis into bins before counting the number of observation in each bin and either represent it as bars (`geom_histogram`) or lines (`geom_freqpoly()`).
- Distribution:
  - `geom_density()`: build the density plot;
  - `geom_boxplot()`: build the well-known boxplot;

[13] This function can also be used to draw arrows through the parameter `arrow` and the function of that same name `arrow()`.

  - `geom_violin()`: application of `geom_density()` displayed in `geom_boxplot()` fashion.
- Text and Labels:
  - `geom_text` and `geom_label`: add text to the graph (see example Section 5.4.2);
  - the package {ggrepel} provides alternative functions (`geom_text_repel()` and `geom_label_repel()`) that reposition labels to avoid overlapping (*repel* stands for *repulsive*).
- Rectangles[14]:
  - `geom_tile()`, `geom_rect`: create area either using its center point (`geom_tile()`) or its four corner (`geom_rect()`) defined by `xmin`, `xmax`, `ymin`, and `ymax`;
  - `geom_raster()`: high-performance alternative to `geom_tile()`/ `geom_rect` where all the tiles have the same size.

Besides `geom_*()`, a lot of graphical parameters can further be controlled. This includes of course the `theme()` and the `aes()`:

- For predefined themes, see example;
- `axis` parameters including their title (`axis.title`), text (`axis.text`), ticks (`axis.ticks`), line (`axis.line`), and all sublevels.
- `legend` parameters including their position (`legend.position`), direction (`legend.direction`), text (`legend.text`, `legend.title`), design of the box (`legend.box`, `legend.background`), etc.
- `panel` parameters including their background (`panel.background`), grid lines (`panel.grid`), border (`panel.border`), etc.
- `plot` parameters including different titles (`plot.title`, `plot.subtitle`, `plot.caption`), background (`plot.backgorund`), etc.

Most of these parameters can be controlled at different levels of granularity:

- overall (e.g. `panel.grid`);
- more detailed (e.g. `panel.grid.major` and `panel.grid.minor`);
- most detailed (e.g. `panel.grid.major.x`, `panel.grid.major.y`, etc).

Depending on whether the option to modify is some text, a line, or a rectangle, `element_text()`, `element_line()`, or `element_rect()` would

[14] In sensory and consumer science, this will often be used for building surface plot responses (e.g. external preference map), and hence is associated to `geom_contour()` to show the different lines.

be, respectively, used to control them. These functions provide general (e.g. `color`) as well as specific options (e.g. `family` and `face` for text, `linetype` for lines) to each type.

Note that if some elements should be left blank, `element_blank()` can be used regardless of the nature of the element.

Let's illustrate these concepts using our previous graph stored in `line_p`. Here, the goal is to remove the grid line, to replace the x and y axis lines by arrows, and to reposition the axis titles to the far end of the axis so that it is next to the arrow head (Figure 5.11).

```
line_p +
  theme(panel.grid=element_blank(),
        panel.border=element_blank(),
        axis.line=element_line(arrow = arrow(ends = "last",
                                             type = "closed")),
        axis.title=element_text(hjust=1))
```

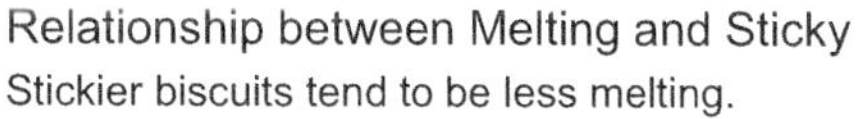


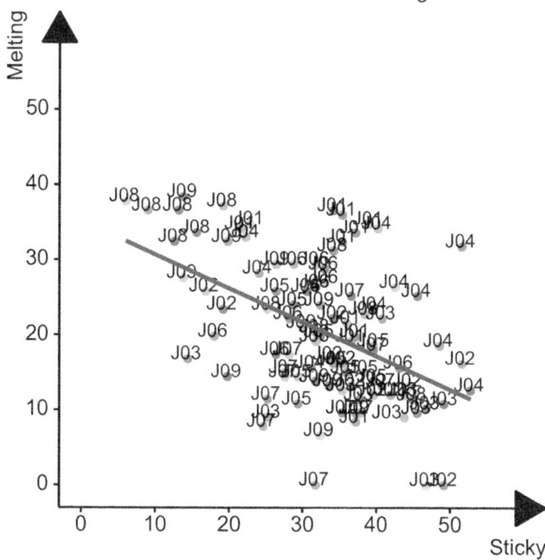


**FIGURE 5.11**
Scatter plot with emphasis on the display through `theme()`.

Similarly to the theme, aesthetics can also be adjusted. In previous examples, the x-axis in `bar_p` was adjusted by setting limits, providing breaks, and replacing the values by certain labels using `scale_x_continuous()`. Most aesthetic parameters can be controlled by equivalent functions for which the name is using the following structure `scale_*nameaes*_*typescale*`, where:

- *nameaes* corresponds to any aesthetics including `x`, `y`, `colour` or `fill`, `alpha`, etc.
- *typescale* corresponds to the type of scale, where it is `continuous`, `discrete`, or `manual` (among others).

Such functions fully control how the corresponding aesthetic should behave, by providing the correspondence between a variable level and (say) its color. In the graph saved in `bar_p`, remember that we filled in the bar chart using the product information. Let's imagine that we are particularly interested in biscuit `P3` and want to compare it to the rest of the biscuits. We propose to make `P3` stand out by filling it in orange and setting all the other biscuits in the same gray tone (Figure 5.12).

Such procedure can be done using `scale_fill_manual()`.

```
bar_p +
  scale_fill_manual(values=c("P1"="gray50", "P2"="gray50",
                             "P3"="darkorange", "P4"="gray50",
                             "P5"="gray50", "P6"="gray50",
                             "P7"="gray50", "P8"="gray50",
                             "P9"="gray50", "P10"="gray50"))
```

When multiple aesthetics are being used, the legend might become overwhelming or redundant. It is possible to turn off some of these visuals within the `scale_*()` functions or by using `guides()` and by setting `nameaes='none'` as shown in the `line_p` example.

### 5.4.4 Miscealleneous

#### Structuring the Axis

By default, `ggplot()` generates plot that fits the data and that fits within the output screen. This means that some graphics might not be perfectly representing the data due to some distortion. In a previous example (`line_p`), the dimensions were made comparable through `coord_fixed()`.

Other transformations can be performed. For instance, the graphic can be transposed using `coord_flip()` as in the following example (Figure 5.13):

```
bar_p + coord_flip()
```

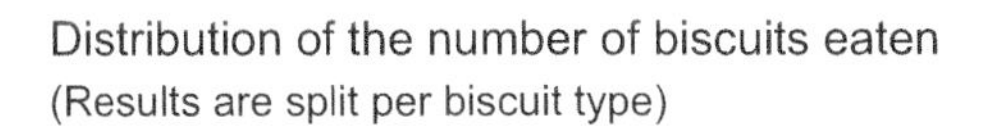


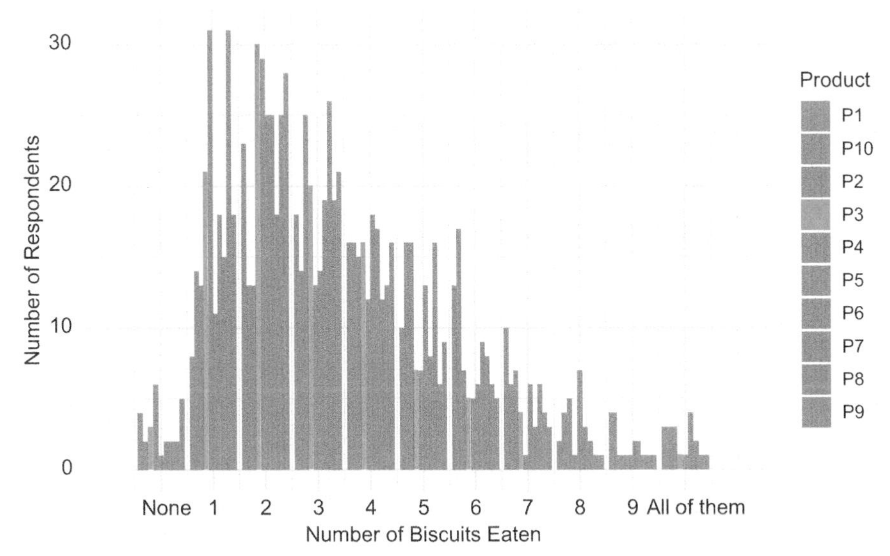


**FIGURE 5.12**
Bar chart with emphasis on the display through `scale_fill_manual()`.

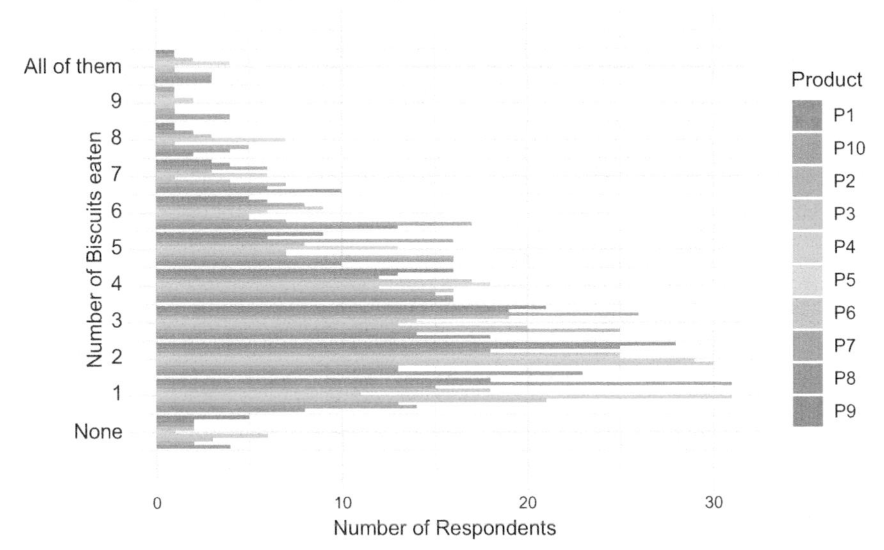


**FIGURE 5.13**
Bar chart with emphasis on the display through `coord_flip()`.

### *Summary Through an Example: Spider Plots*

To conclude this section, and summarize most concepts presented in this chapter, let's introduce the well-known spider plots. The use of such plots are quite polarizing among analysts and the reason of this choice here is purely educational, as **1.** there are no predefined options in {ggplot2} that provide such charts and **2.** they present some interesting challenges.

Let's start with deconstructing a spider-plot: a spider plot is a line chart presented in a circular way. So let's start with building a line chart of our sensory profiles (the means are considered here). For more clarity, only two of the samples (P03 and POpt) are represented.

```
sensory_mean <- sensory %>%
  pivot_longer(Shiny:Melting,
               names_to="Variables", values_to="Scores") %>%
  mutate(Variables = fct_inorder(Variables)) %>%
  group_by(Product, Variables) %>%
  summarize(Mean = mean(Scores)) %>%
  ungroup() %>%
  filter(Product %in% c("P03", "POpt"))

spider_line <- sensory_mean %>%
  ggplot(aes(x=Variables, y=Mean, colour=Product, linetype=Product))+
  geom_point(pch=20, cex=3)+
  geom_line(aes(group=Product), lwd=1)+
  theme_minimal()+
  xlab("")+
  scale_y_continuous(name="", labels=NULL, limits=c(0,50))+
  scale_colour_manual(values=c("P03"="darkorange", "POpt"="grey50"))+
  scale_linetype_manual(values=c("P03"="solid", "POpt"="dashed"))
```

Next step is to represent this line chart in a circular way (Figure 5.14). This can be done using coord_polar():

```
spider_line + coord_polar()
```

This already looks like a spider plot! However, a closer look at it highlights a point that needs improvement: There is no connection between the last attribute (Melting) and the first one (Shiny).

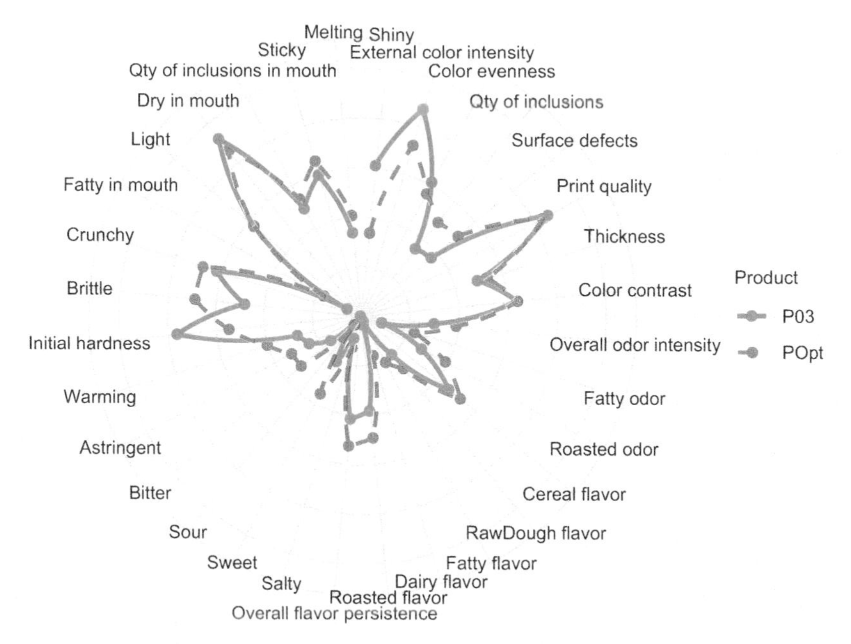


**FIGURE 5.14**
Line charts using `coord_polar()`.

To counter this, the following two-fold solution is proposed:

1. Associate each attribute to its position (e.g. `Shiny` is 1, `External color intensity` is 2, until `Melting` which would be 32);
2. Duplicate the last attribute (`Melting`) and associate it to position 0.

```
var <- levels(sensory_mean$Variables)

sensory_mean_pos <- tibble(Variables = c(var[length(var)], var),
                           Position = 0:length(var)) %>%
  full_join(sensory_mean, var_pos, by="Variables")
```

The previous graph is then rebuilt by forcing the position of the attributes on the x-axis using `Position` (`Variables` is used for the labels) (Figure 5.15). Here position 0 is voluntarily omitted (`breaks = 1:length(var)` and `labels = var`), meaning that only the labels going from 1 to the last variable are being showed. However, the x-axis is forced to go from 0 to the number of attributes (`limits = c(0, length(var))`).

```
spider_plot <- sensory_mean_pos %>%
  ggplot(aes(x=Position, y=Mean, colour=Product, linetype=Product))+
  geom_point(pch=20, cex=2)+
  geom_line(aes(group=Product), lwd=1)+
  theme_minimal()+
  scale_x_continuous(name="", breaks=1:length(var),
                     labels=var, limits=c(0,length(var)))+
  scale_y_continuous(name="", labels=NULL, limits=c(0,50))+
  scale_colour_manual(values=c("P03"="darkorange", "POpt"="grey50"))+
  scale_linetype_manual(values=c("P03"="solid", "POpt"="dashed"))+
  coord_polar()+
  theme(legend.position = "bottom", legend.title = element_blank())
```

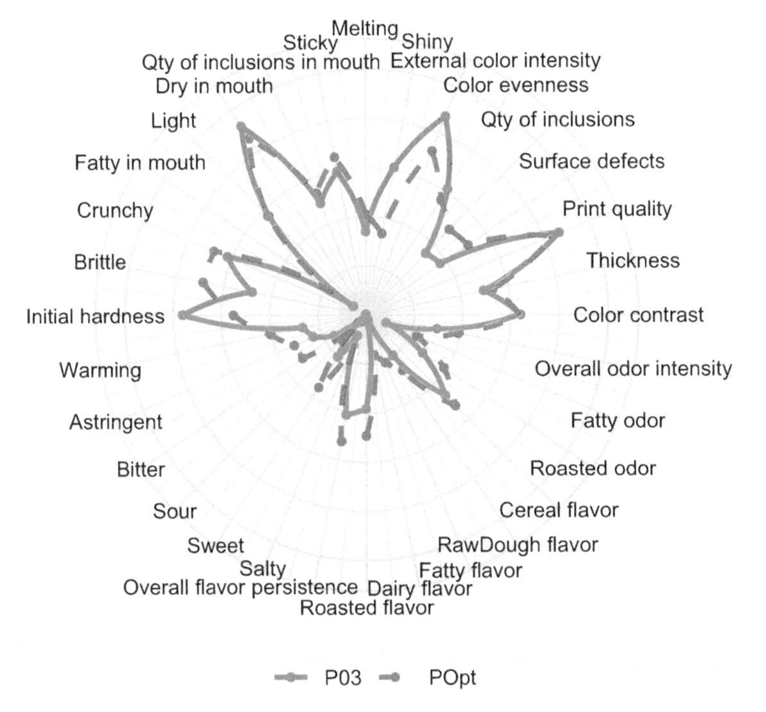


**FIGURE 5.15**
Spider plot.

By using this trick, the connection between the first and last attributes is established.

## *Combining Plots*

When multiple plots should be generated using the same pattern on subset of data, it is possible to generate them automatically using `facet_wrap()`

or facet_grid(). The difference between these two functions rely on the number of variables to use for the split: In facet_wrap(), the graphics are *vectorized*, meaning that each element of the split is represented independently. For facet_grid(), however, the graphics is represented in a matrix, meaning that two blocks of split variables are required, one for the columns and one for the rows.

An example of facet_wrap() is provided in Section 10.2.2.

For these two functions, the parameter scales is particularly interesting as it allows each separate graph to use its own axis scales (free or individually using free_x/free_y) or not (fixed).

To go further, consider also the function facet_trelliscope() from the {trelliscopejs} package. This function generates the same type of graphs as facet_wrap() or facet_grid() with some powerful twists. After generating the plots, they are still editable thanks to an interactive menu that controls for the *Grid*, *Labels*, *Filter*, and *Sort*. For example, the number of plots to show per row/column can be adjusted and tables with descriptive statistics (e.g. mean, minimum, maximum) can be added under each graph, etc. Moreover, readers that are familiar with the interactivity of {plotly} can make great use of it through the as_plotly = TRUE parameter, which is then applied to each individual graph!

Such procedure is very handy to produce multiple graphs all at once... when the data allow it. When multiple plots are being generated separately (using different data set or producing different types of plots), it can still be relevant to combine them all in one. To perform such collage, the package {patchwork} becomes handy.

{patchwork} is a package that allows combining ggplot() graphs using *mathematical* operations. To add two elements next to each others, the + sign is used. To add two elements on top of each others, they should be separated using /. This operation can be combined with () to generate fancier collage.

Let's illustrate this by creating a plot with spider_plot on the left side and bar_p on the right side on top of line_p.

```
library(patchwork)

p <- spider_plot + (bar_p / line_p)
```

A general title can be added, as well as tag levels (handy for publications!), using plot_annotation() (Figure 5.16).

```
p + plot_annotation(title = "Example of 'ggplots' I've learned today",
                    tag_levels='a')
```

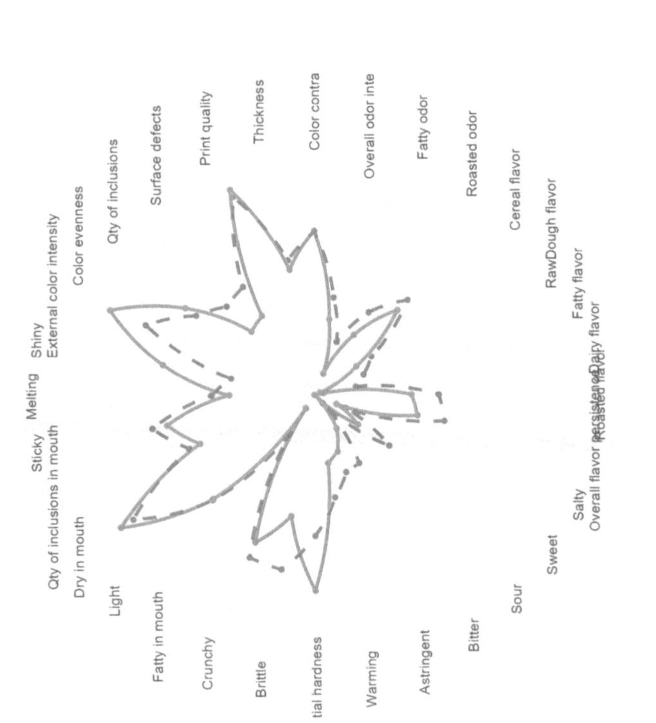


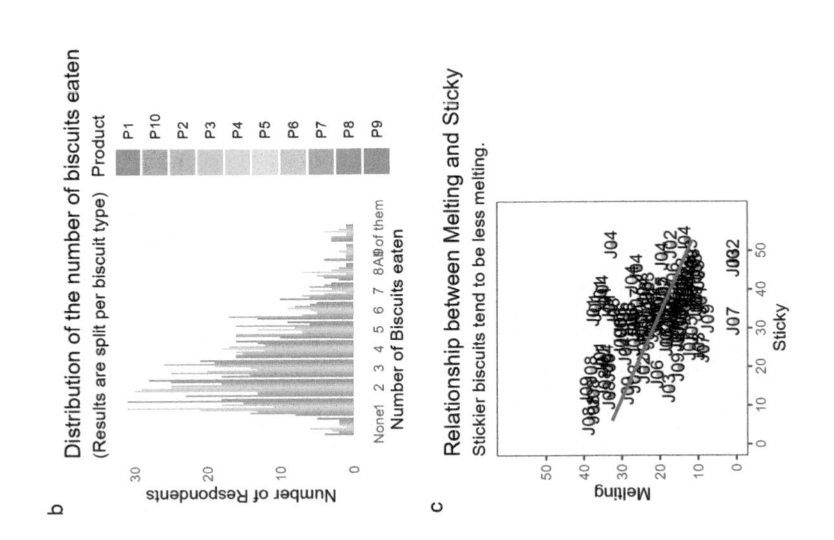


**FIGURE 5.16**
Multiple plot examples.

### 5.4.5 Few Additional Tips and Tricks

#### Combining Data Transformation and {ggplot2} Grammar

Both {tidyverse} and {ggplot2} use pipes to combine lines of code or layers. However, the pipes themselves are defined differently since {maggritr} uses %>% whereas {ggplot2} uses +. It is, however, possible to combine both systems one after each other, but just remember to switch from %>% to + as you transition from data transformation/tidying to building your graph (see Chapter 4).

#### Ordering Elements in a Plot

When building a graph using categorical variables, {ggplot2} tends to represent the different levels in alphabetical order, especially if the variable is defined as character. Such situations can make the graph more difficult to read, as the categories may not be presented in a logical order (e.g. fall, spring, summer, and winter instead of spring, summer, fall, and winter). To ensure that the elements are in the right order, either consider transforming the variables into factors (using factor() by indicating the levels order of your choice or through fct_inorder() to keep the order from the file, Chapter 4) or by using a position variable as in the spider_plot example. The former option also works for ordering elements in the legend.

If the order of the elements should be changed *within the charts* (simple changes such as reverting the order), it can be done directly within the geom_*() function. This is, for instance, the case with the stacked bar chart, in which the order may be reverted using the parameter position = position_fill(reverse = TRUE) (suggesting here that the split was defined through fill in aes()).

#### Fixing Overlapping Axis Text

When ggplot() are being built using categorical variables, the labels used on the x-axis are often overlapping (some of the labels being then unreadable). A first good/easy solution consists of reducing the size of the label and/or shortening them as long as it does not affect its readability. However, this might not always be possible or sufficient, and other adjustments are required. Let's use spider_line as illustration to show three possible solutions.

The first option consists of using theme() and rotating the labels (here at 45 degrees, but use 90 degrees to get the names vertically) (Figure 5.17). Note that by default, ggplot() centers the labels: to avoid having them crossing the x-axis line, they are being left-centered using hjust=1:

```
spider_line +
  theme(axis.text.x = element_text(angle=45, hjust=1))
```

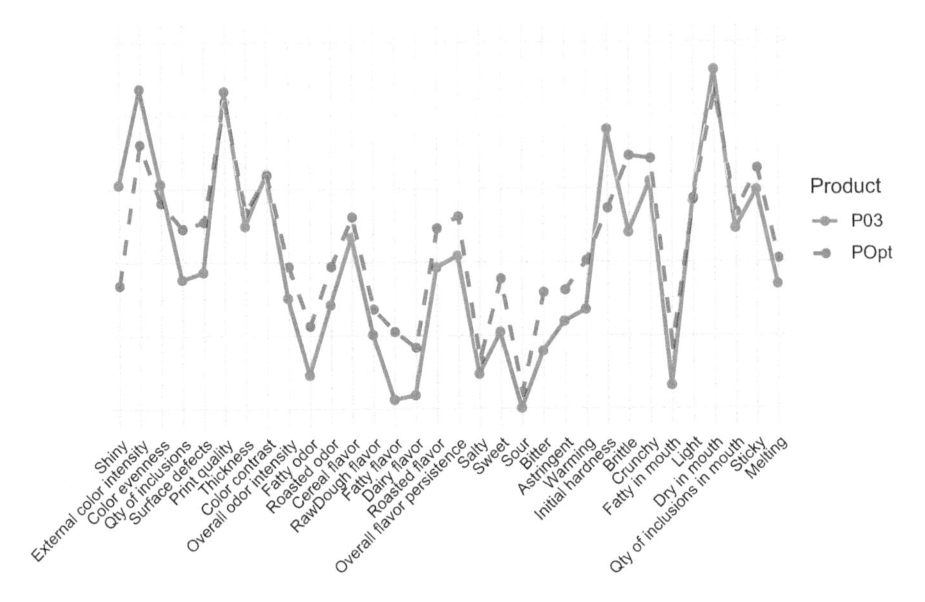


**FIGURE 5.17**
Spider plot fixing overlapping through `theme ()`.

A second option consists in dodging one every two labels along the X-axis (Figure 5.18). This option works fine, especially when the labels are not too long. In our example, unfortunately, some overlap can still be seen. Note that this option is accessible within `scale_x_discrete()`, but not within `theme()` as we would expect:

```
spider_line +
  scale_x_discrete(guide = guide_axis(n.dodge = 2))
```

The last option consists of transposing the graph using `coord_flip()`. This solution works well since labels on the y-axis are written horizontally. However, this option is not always suitable due to conventions: If it is recommended for bar charts, it may not be for line charts, for instance.

### Exporting Graphs

There are various ways to save or export `ggplot()` charts. To save these plots to your computer in various formats (e.g. png, pdf, etc.), `ggsave()` is used. By default, `ggsave()` exports the last plot built and saves it in the location defined by `filename`, in the format defined by `device` (additional information regarding the width, height, dpi, etc. can also be configured).

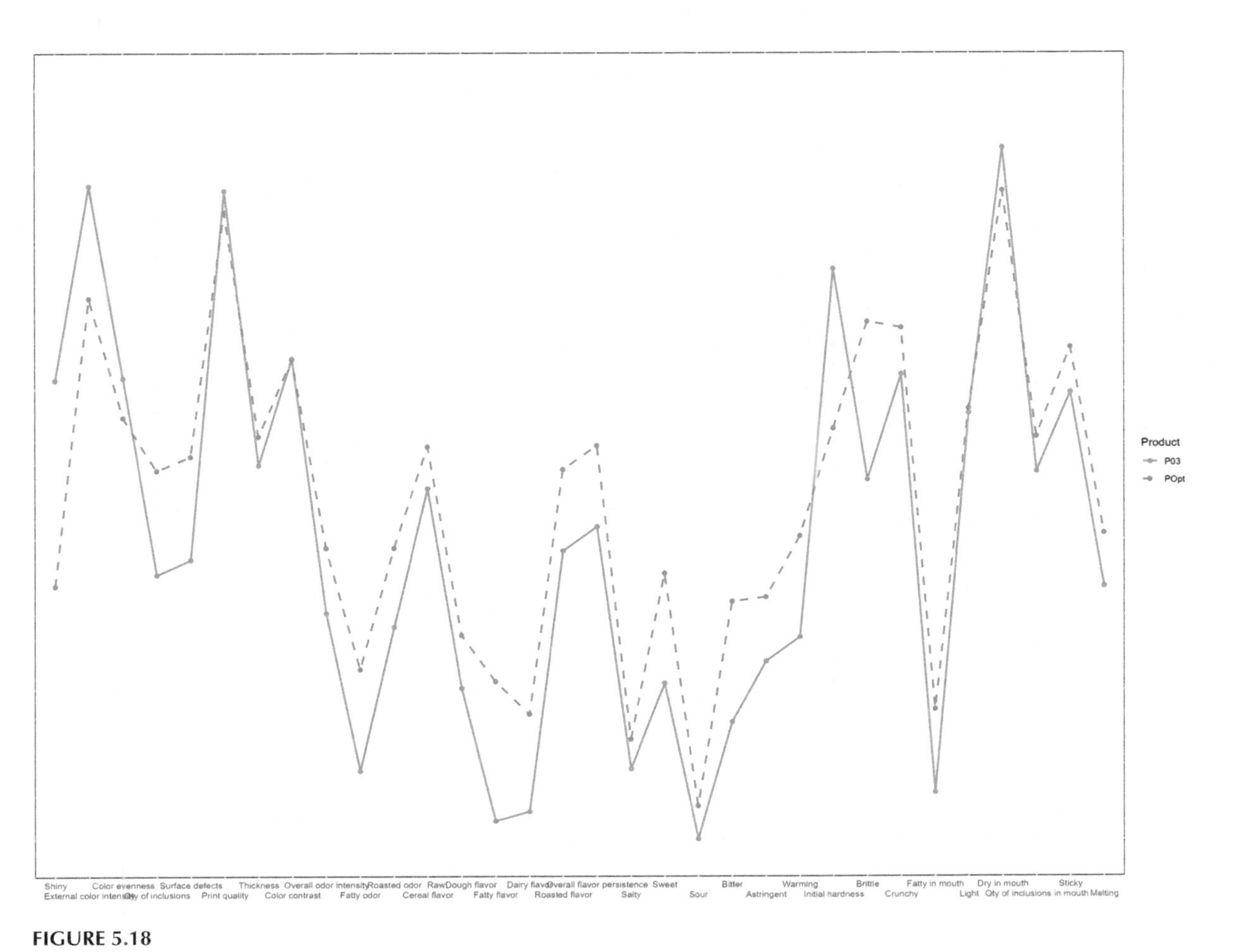


**FIGURE 5.18**
Spider plot fixing overlapping through `scale_x_discrete()`.

For instance, the spiderplot generated earlier[15] can be saved as follows:

```
ggsave(filename="spiderplot.png", plot=spiderplot, device="png")
```

As an alternative, ggplot() graphs can also be exported in PowerPoint or Word through the {rvg} package (see Section 6.2.2).

### *Additional Libraries*

{ggplot2} is a very powerful tool for data visualization. By default, it offers a very large variety of possibilities, which should cover most situations that you would encounter. If not, a quick internet search will most likely find extensions in alternative packages that will provide you with solutions.

To help you further, here is a nonexhaustive list of relevant packages:

- {ggcharts}: This package provides nice and clear alternatives to some {ggplot2} options through simple functions in one line of code, including bar_chart(), line_chart(), lollipop_chart(), and dumbbell_chart() just to name a few.
- {graffify}: This package extends {ggplot2} by providing nice and easy functions to help data visualization and linear models for ANOVA. For example, it generates one function bar chart with error bars through plot_scatterbar_sd() or simultaneous box-plot and scatter plot through plot_scatterbox().
- {factoextra}: Although {FactoMineR} generates its graphs in {ggplot2} and in base R, {factoextra} is a great extension as it is easy to use and provides a wide variety of options to customize your plots.
- {ggcorrplot}: There are many packages that propose to visualize graphically tables of correlations. However, we particularly like this one for its simplicity.
- {ggwordcloud}: It is a great package for building word-clouds as it provides a large degree of control. With this package, the words can either be positioned randomly or by matching a predefined shape, etc. But more interestingly, words can also be positioned semi-randomly, hence giving more interpretation power to the final results (for more information, please visit (see *ggwordcloud*).[16]

[15] Since spiderplot is not the last plot generated, it needs to be defined in plot.
[16] https://lepennec.github.io/ggwordcloud/

- {ggraph}: This package provides neat solutions to build network visualization in {ggplot2}.
- {performance}: This package provides predefined graphs that allow you to evaluate the quality of your models through the single check_model() function. See also {ggside} if you want to print on the margin of your regression plot the marginal distributions (or density plot) of your different categories.

To learn more about {ggplot2} basics, we recommend two additional sources of information:

- {esquisse}: After loading the package, run the function esquisser(). This command opens a window in which you can select your data set (the data set should be available within you R environment), the type of plot to build, and all the relevant information to build your plot (which variable to be used as X-axis, Y-axis, etc.) through an user-friendly interface. Ultimately, the graph is being generated, but more importantly, the code used to generate the plot is provided. This is hence an educational tool to learn build graphs with {ggplot2}.
- *from Data to Viz*[17] provides a wide gallery of graphics sorted by the type of data that you have. Each graphic proposed is illustrated with an example provided in R (often in {ggplot2}) and in Python. This website is hence inspirational and educational both at the same time!

[17] https://www.data-to-viz.com/

# 6

# *Automated Reporting*

Learning a programming language is not only useful for the freedom it provides, it also increases largely the speed of the analysis thanks to the reusability of the code. When many projects are built using the same base (e.g. similar questionnaires), the same analyses are often performed, also leading to similar reports. Automating the analysis process is simple by applying the same code to different data sets. But what about the reports? Should a new report be built manually for each project? In this section, we will show you how to build your own report directly from your code next to your analyses (or integrate your analysis within your reporting process). Such procedure has three main benefits: 1) build a new standardized report automatically for each new data set analyzed, 2) mass-produce slides automatically, and 3) save time for the interpretation of the results and the storytelling.

## 6.1 What and Why Automated Reporting?

Effective communication of results is among the essential duties of sensory scientists... and so is data collection, data preparation, data analysis, etc. Worst, the sometimes tedious mechanics of report production together with the sheer volume of data that many scientists must process combine to make reporting design and nice storytelling an afterthought in too many cases. Although this should preferably not be happening, it is necessary sometimes as presentation deadlines approach and time becomes limited. Add to this work-load some last-minute changes due to either a change in the data, in the analysis, or maybe an error (e.g. copy/paste the wrong column, etc.) you have just detected... how can we be sure that the latest version of the report is fully up-to-date and that all the results/conclusions are correct? As the statistical software (e.g. R) is often separated from the reporting tool (e.g. Microsoft Office), it is easy to miss to transfer updated outputs (e.g. values, tables, figures, etc.) to the report, hence creating inconsistencies.

DOI: 10.1201/9781003028611-6

Let's consider another scenario, in which all the preliminary steps have been successfully done. After a week of analyses and reporting, you present your report to your manager or clients and they come with a question such as: "Can we deep-dive into the results by looking at a particular group of consumers (e.g. gender split, or cluster split)?" Do you feel like going through the entire process again?

How would you feel if we would tell you that there is a way to build your report while running the analysis, by using the same script file? This means that in few clicks, say after updating the data to analyze (e.g. filter to the target group of consumers only), your report gets automatically regenerated with all the newly updated results. Such solution seems ideal since it increases efficiency while reducing errors due to manual processing of results. More importantly, this gain in time and effort allows you designing nicer slides and building a better story.

## 6.2 Integrating Reports within Analysis Scripts

In this section, let's integrate our report building process within our data analysis. By doing so, we do not focus on building a story yet. Instead, we improve our way of working by exporting directly all the statistical outputs that could[1] be useful for our future storytelling. By doing so, we increase efficiency (especially if code can be reused for other studies) by *killing two birds with one stone*: We simultaneously run our analysis, create usable content for our final presentation, while reducing errors due to manual processing.

Since Microsoft Office is often the tool used for sharing results, we will focus our attention in exporting results to Excel, PowerPoint, and Word.

As usual, let's start by loading the general package that we would need for our analyses (more specific packages being mentioned later.)

```
library(tidyverse)
library(here)
library(readxl)
```

*Note that in this chapter on automated reporting, some results (tables and figures) that are being created in one of the section may be reused in subsequent sections. In case you do not read this chapter linearly, you might get errors as*

[1] We say *could* as we are in a process of mass-exportation of results, most of them being used for building the story although they may not be kept in the final deck.

*you might be calling objects that do not exist yet in your environment. If that should be the case, read through the previous sections to find the code where these elements are being generated, run it, and resume your read.*

### 6.2.1 Excel

Although Excel is not our preferred tool for automated reporting, it is still one of the major ways to access and share data. Most data collection software offer the possibility to export data and results in Excel, while most data analysis software accept Excel format as inputs. With the large use of Excel, it is no surprise that many colleagues and/or clients like to share data and results using spreadsheets. It is even less a surprise that R provides multiple solutions to import/export results from/to Excel.

For importing Excel files, we have already presented the package {`readxl`} among others (see Section 8.4). For exporting results, two complementary packages (yet again, among others!) in terms of ease of use and flexibility in the outcome are proposed: {`writexl`} and {`openxlsx`}.

As its name suggests, {`writexl`} is dedicated to exporting tables to Excel through the `write_xlsx()` function. Its use is very simple as it only takes as inputs the table (or list of tables)[2] to export to the file specified in the `path` parameter.

Let's illustrate this by using our *biscuits_sensory_profile.xlsx* file: Let's imagine that we would like to reduce our data set by only considering products that are high in Protein:

```
file_path <- file.path("data", "biscuits_sensory_profile.xlsx")
product_info <- readxl::read_xlsx(path=file_path, sheet="Product Info",
                                  range="A1:D12", col_names=TRUE)

high_prot <- product_info %>%
  filter(Protein %in% "High") %>%
  pull(Product)

high_prot_data <- readxl::read_xlsx(path=file_path, sheet="Data") %>%
  filter(Product %in% high_prot)
```

We then export this data into an excel sheet called *export.xlsx* that will be contained in our folder called *output*[3]:

---

[2] List of tables will generate multiple sheets within the same spreadsheet, one table being placed in each sheet.

[3] If the *output* folder does not exist, this code will return an error so make sure to create one.

```
library(writexl)
write_xlsx(high_prot_data, path="output/export.xlsx", col_names=TRUE)
```

The export of tables using {writexl} is intuitive and easy, yet simplistic as it does not allow formatting the tables (except for some minor possibilities for the header) nor does it allow exporting multiple tables within the same sheet. For more advanced exporting options, the use of {openxlsx} is preferred as it allows more flexibility in structuring and formatting the Excel output.

With {openxlsx}, the procedure starts with creating a workbook object (e.g. `wb`) using `createWorkbook()`. We can add worksheets to `wb` through `addWorksheet()`.

```
library(openxlsx)
wb <- openxlsx::createWorkbook()
addWorksheet(wb, sheetName = "Mean", gridLines = FALSE)
```

Note that with `addWorksheet()`, it is possible to control the appearance of the worksheet:

- show/hide grid lines using `gridLines`;
- color the sheet using `tabColour`;
- change the zoom on the sheet through `zoom`;
- show/hide the tab using `visible`;
- format the worksheet by specifying its size (`paperSize`) and orientation (`orientation`).

On a given worksheet, any table can be exported using `writeData()` or `writeDataTable()`, which controls where to write the table through the `startRow` and `startCol` options.

Let's imagine we want to compute the sensory profiles of the products, and we want to export that into Excel. Rather than simply exporting the results, we want to customize the output by applying the Excel style named *TabelStyleLight9*:

```
# Creating the Sensory Profiles with some Product Information
p_info <- readxl::read_xlsx(file_path, sheet = "Product Info") %>%
  dplyr::select(-Type)
sensory <- readxl::read_xlsx(file_path, sheet="Data") %>%
```

```
  inner_join(p_info, by="Product") %>%
  relocate(Protein:Fiber, .after=Product)
senso_mean <- sensory %>%
  pivot_longer(Shiny:Melting,
               names_to="Attribute", values_to="Score") %>%
  dplyr::select(-Judge) %>%
  pivot_wider(names_from=Attribute, values_from=Score,
              values_fn=mean)

# Add the table to the Excel Sheet
writeDataTable(wb, sheet="Mean", x=senso_mean,
               startCol=1, startRow=1,
               colNames=TRUE, rowNames=FALSE,
               tableStyle="TableStyleLight9")
```

At any time, you can visualize the Excel file that is being produced without exporting it yet using `openXL()`. This function comes very handy as it allows you checking that the output looks like what you would wish for.

```
openXL(wb)
```

As can be seen, `writeData()` and `writeDataTable()` give us a lot of control on our export. For instance, we can:

- control where to print the data by using `startRow` and `startCol` (or alternatively `xy`: `xy = c("B",12)` prints the table starting in cell B12), hence allowing exporting multiple tables within the same sheet;
- include the row names and column names through `rowNames` and `colNames`;
- format the header using `headerStyle` (incl. color of the text and/or background, font, font size, etc.);
- apply a specific style to our table using `tableStyle`;
- shape the borders using predefined solutions through `borders`, or customizing them with `borderStyle` and `borderColour`;
- add a filter to the table using `withFilter`;
- convert missing data to "#N/A" or any other string using `keepNA` and `na.string`.

Rather than using some predefined formatting as was the case with `tableStyle`, let's consider some more advanced options in which we control (almost) everything. Let's start with setting up the formatting style we would like to apply:

```
# Pre-define options to control the borders
options("openxlsx.borderColour" = "#4F80BD")
options("openxlsx.borderStyle" = "thin")

# Automatically set Number formats to 1 value after the decimal
options("openxlsx.numFmt" = "0.0")

# Change the font to Calibri size 10
modifyBaseFont(wb,fontName = "Calibri", fontSize = 10)

# Header Style (blue background, top/bottom borders, text centered/bold)
headSty <- createStyle(fgFill="#DCE6F1", border="TopBottom",
                       halign="center", textDecoration="bold")
```

Note that many more formatting options can be configured through:

- `options()` to predefine number formatting, border colors and style, etc.;
- `modifyBaseFont()` to define the font name and font size;
- `freezePane()` to freeze the first row and/or column of the table;
- `createStyle()` to predefine a style or `addStyle()` to apply the styling to selected cells;
- `setColWidths()` to control column width;
- `conditionalFormatting()` to format cells based on predefined conditions.

Let's export again the sensory profiles in a second sheet after applying these formatting options:

```
addWorksheet(wb, sheetName = "Mean (manual formatting)", gridLines = FALSE)

freezePane(wb, sheet=2, firstRow=TRUE, firstCol=TRUE)

writeData(wb, sheet=2, x=senso_mean,
          startCol=1, startRow=1,
          colNames=TRUE, rowNames=FALSE,
          headerStyle=headSty)
```

You'll notice that the same table is now presented in a different way (use `openXL(wb)` to view it).

Let's now consider a third export of the sensory profiles, with an additional twist: for a given variable (i.e. column), the value is colored in red (resp. blue) if it is higher (resp. lower) than its mean. To do so, we need to use conditional formatting.

Let's start with creating two predefined parameters called `pos_style` (red) and `neg_style` (blue) using `createStyle()` that we will use to color the different cells. Let's also compute the overall mean per attribute.

```
# Styles for conditional formatting
pos_style <- createStyle(fontColour = "firebrick3",
                         bgFill = "mistyrose1")
neg_style <- createStyle(fontColour = "navy",
                         bgFill = "lightsteelblue")

# Compute the overall mean
overall_mean <- senso_mean %>%
  summarize(across(where(is.numeric), mean))
```

Let's then create a new worksheet in which we print the data of interest:

```
addWorksheet(wb, sheetName = "Conditional Formatting", gridLines=FALSE)
writeDataTable(wb, sheet=3, x=senso_mean,
               startCol=1, startRow=1, colNames=TRUE, rowNames=FALSE)
```

Finally, we color the cells according to the rules that were defined earlier. To do so, the decision whether `pos_style` or `neg_style` should be used is defined by the `rule` parameter from the `conditionalFormatting()`[4] function.

```
# Adding formatting to the second column
for (v in 1:ncol(overall_mean)){
  conditionalFormatting(wb, sheet=3, cols=v+3,
                        rows=1+1:nrow(senso_mean),
                        rule=paste0(">", overall_mean[1,v]),
                        style=pos_style)
  conditionalFormatting(wb, sheet=3, cols=v+3,
                        rows=1+1:nrow(senso_mean),
                        rule=paste0("<", overall_mean[1,v]),
                        style=neg_style)
}
```

Few comments regarding this code:

- We want to run this for each sensory attribute, hence the `for` loop that goes from 1 to the number of columns stored in `overall_mean`

[4] In `conditionalFormatting()`, you can specify to which `rows` and `cols` the formatting applies.

(`overall_mean` only contains the overall mean scores for the sensory attributes);

- `senso_mean` however contains three extra columns: `Product`, `Protein`, and `Fiber` hence the parameter `cols = v + 3`;
- We apply the formatting to all the rows except the header, hence `rows = 1 + 1:nrow(senso_mean)`;
- Finally, we apply `pos_style` (resp. `neg_style`) if the value is larger (resp. lower) than the overall mean for that attribute using `rule = paste0(">", overall_mean[1,v])` (resp. `rule = paste0("<", overall_mean[1,v])`).

Once the spreadsheet is complete, we export the results using `saveWorkbook()` by specifying the name of the workbook (`wb`) and its path through `file`. In case such workbook already exists, it can be overwritten using `overwrite=TRUE`.

```
saveWorkbook(wb, file="output/export2.xlsx")
```

For more information regarding {`openxlsx`}, please visit https://rdrr.io/cran/openxlsx/.

### 6.2.2 PowerPoint

#### *Creating a PowerPoint Deck*

Throughout the years, PowerPoint became one of the main supports for presenting results, whether it is in academia, in conference, or in companies. It is hence important to show how reports can be generated in PowerPoint from R. Many solutions exist, however, the {`officer`} package is used here as its application is vast while still remaining easy to use.

---

{`officer`} contains a conflicting function with {`readxl`} in `read_xlsx()`. To ensure you use the right function, call the function from the package of interest (e.g. `readxl::read_xlsx()`).]

---

With {`officer`}, the procedure starts with creating a PowerPoint object (`pptx_obj`) using the `read_pptx()` function.

```
library(officer)
pptx_obj <- read_pptx()
```

A blank deck is set up with the *Office Theme*. To use a custom theme and custom slides, a predefined deck from PowerPoint software can be used as input. Let's import the *example.pptx* template that we created for you:

```
pptx_obj_custom <- read_pptx(file.path("data", "example.pptx"))
```

The content of the template can be inspected through `layout_summary()`:

```
pptx_obj %>%
  layout_summary()
```

```
##                 layout       master
## 1        Title Slide Office Theme
## 2 Title and Content Office Theme
## 3     Section Header Office Theme
## 4        Two Content Office Theme
## 5         Comparison Office Theme
## 6         Title Only Office Theme
## 7              Blank Office Theme
```

As can be seen by `layout_summary()`, the default template imported (also called *master*, which is defined here as *Office Theme*) proposes seven types of slides including *Title Slide*, *Title and Content*, *Section Header*, *Two Content*, *Comparison*, *Title Only*, and finally *Blank*. The *example.pptx* template has 11 different types of slides and contains custom *master slides* called *Integral*.[5]

Each of these slides present some predefined properties (e.g. a box for text of tables/images, a header, etc.). Let's look at the properties of *Title and Content* using `layout_properties()`:

```
pptx_obj %>%
  layout_properties() %>%
  filter(name == "Title and Content") %>%
  as_tibble()
```

```
## # A tibble: 5 x 13
##   master_~1 name  type  id    ph_la~2 ph     offx  offy
##   <chr>     <chr> <chr> <chr> <chr>   <chr> <dbl> <dbl>
```

[5] It is out of the scope of this book to describe how to build your own custom master slides. However, a quick internet search will provide you with all the information that you would need.

```
## 1 Office T~ Titl~ body  3     Conten~ "<p:~  0.5   1.75
## 2 Office T~ Titl~ dt    4     Date P~ "<p:~  0.5   6.95
## 3 Office T~ Titl~ ftr   5     Footer~ "<p:~  3.42  6.95
## 4 Office T~ Titl~ sldN~ 6     Slide ~ "<p:~  7.17  6.95
## # ... with 1 more row, 5 more variables: cx <dbl>,
## #   cy <dbl>, rotation <dbl>, fld_id <chr>,
## #   fld_type <chr>, and abbreviated variable names
## #   1: master_name, 2: ph_label
```

This code provides more details about the elements available in each type of slides, including their identifiers and positions on the slide. This information is required to export content in some specific elements.

---

Unfortunately, {officer} does not provide a function similar to `openxlsx::openXL()` that allows visualizing the file that is currently being built. Instead, the document needs to be saved on the disk using the `print()` function, which takes as entries the PowerPoint file to export (here `pptx_obj`) and its output location.

---

### Adding/Moving/Removing Slides

With {officer}, various actions can be done on the slides. The first logical action consists in adding a new slide to a presentation, in which we will later on export some text, tables, figures, etc. Such action can be done using `add_slide()`, in which the type of slide and the master[6] are informed:

```
master <- "Office Theme"
pptx_obj <- pptx_obj %>%
  add_slide(layout = 'Title and Content', master = master)
```

This code adds slide of type *Title and Content* to your deck.

Additional operations on the slides themselves can be done. In particular, you can reorganize your deck by changing the orders of your slides using `move_slide()`, delete slides that are no longer needed through `remove_slide()`, or modify a preexisting slides by making it active using `on_slide()` (by default, the last slide created is the active one).

For example, let's add another slide of type *Two Content*:

```
pptx_obj <- pptx_obj %>%
  add_slide("Two Content", master=master)
```

[6] In practice, a unique template can contain slides from different masters.

In case we would want to move this slide to eventually be first, the following code is used:

```
pptx_obj <- pptx_obj %>%
  move_slide(index=2, to=1)
```

Ultimately, this slide (now positioned as first slide) can be removed (by default, `index=NULL` and the active slide is deleted):

```
pptx_obj <- pptx_obj %>%
  remove_slide(index=1)
```

### Positioning Information on the Slide

On a given slide, any type of content (text, graph, table, etc.) can be exported. To do so, we need to inform *where* to write *what.*

As we will see in the next sections, the *what* can be any R element including simple text, tables, figures, etc. So let's ignore it for the moment and let's focus on *where.*

To inform where to print elements on the slide, the function `ph_with()` (*ph* stands for *placeholder*) is used. In practice, `ph_with()` comes with the parameter `location`, which takes as input a *placeholder location object* predefined by the function `ph_location()` or one of its derivative, one of the most useful one being `ph_location_type()`. To do so, simply provide the name stored in the column type from the `layout_properties()` output presented before, as following:

```
my_data <- c("My functions are:", "ph_with", "ph_location_type")

pptx_obj <- pptx_obj %>%
  ph_with(value = "My first title",
          location = ph_location_type(type = "title")) %>%
  ph_with(value = my_data,
          location = ph_location_type(type = 'body'))
```

This code adds a title ("My first title") and the text stored in `my_data` to the *body* of the slide (*Title and Content*) created previously.

Other predefined alternatives to `ph_location()` include:

- `ph_location_fullsize()` to produce an output that covers the entire slide;

- `ph_location_left()` and `ph_location_right()` to write in the left/right box in *Two Content* types of slide;
- `ph_location_label()` is similar to `ph_location_type()` except that it uses the label rather than the type.

For a full control of the position where to print your element, `ph_location()` is used as it allows specifying the exact location (for left/top/width/height, units are expressed in inches):

```
my_data2 <- "My new text positioned using ph_location()"

pptx_obj <- pptx_obj %>%
  add_slide(layout = "Title and Content", master = master) %>%
  ph_with(value = my_data2,
          location = ph_location(left=2, top=2, width=3, height=1))
```

To visualize the different steps done so far, let's save the results on our computers in an object called *my export.pptx* stored in the folder *output*:

```
print(pptx_obj, "output/my export.pptx")
```

### Exporting Text

In the previous section, we already exported text to slides. Let's go a bit deeper in the process by also showing how to format the text.

By default, each new text item added to a PowerPoint via {`officer`} is a paragraph object. To further format the paragraph, three main functions are being used:

- `fpar()` (*formatted paragraph*) creates the paragraph;
- `ftext()` (*formatted text*) allows editing the text before pasting into paragraphs. `ftext()` requires a second argument called `prop` which contains the formatting properties;
- `block_list()` allows us to wrap multiple paragraphs together.

Additionally, the text itself can be formated (font, size, color, etc.) using `fp_text()`. Let's go through an example to illustrate the use of these functions:

```
# Formatting option
my_prop <- fp_text(color = "red", font.size = 14)
# First line of text, formatted
my_text <- ftext("First Line in Red", prop = my_prop)

# text into a paragraph
my_par <- fpar(my_text)
# other empty paragraph to introduce an empty line
blank_line <- fpar("")

# second line of text, unformatted
my_par2 <- fpar("Second Line")
# Final block with the two lines of text separated by the empty line
my_list <- block_list(my_par, blank_line, my_par2)

pptx_obj <- pptx_obj %>%
  add_slide(layout = "Title and Content", master = master) %>%
  ph_with(value = my_list,
          location = ph_location_type(type = "body"))
```

Again, if you want to visualize the results, simply print the results as earlier:

```
print(pptx_obj, target = "output/my export.pptx")
```

This adds an additional slide to our previous PowerPoint deck with our formatted text.

The last element of formatting to consider is the hierarchy in bullet points. Let's add a slide containing three bullet points with a hierarchy so that the first and third lines are the primary points and the second line is a secondary point. Such hierarchy is informed using the `level_list` parameter, which informs the hierarchy of each element:

```
text1 <- fpar("FIRST SENTENCE")
text2 <- fpar("second sentence")
text3 <- fpar("THIRD SENTENCE")
my_data <- block_list(text1, text2, text3)

pptx_obj <- pptx_obj %>%
  add_slide(layout = "Title and Content", master = master) %>%
  ph_with(value = my_data, level_list = c(1,2,1),
          location = ph_location_type(type = 'body'))
```

### Exporting Tables

After exporting formatted text to slides, let's export tables.

This can be done by rendering a data frame rather than text as `ph_with()` accepts it and exports it in a default format. Let's use a subset of `senso_mean`[7] for illustration:

```
ft_data <- senso_mean %>%
  dplyr::select(Product, Salty, Sweet, Sour, Bitter) %>%
  mutate(across(where(is.numeric), round, 2))

pptx_obj <- read_pptx() %>%
  add_slide(layout = "Title and Content", master = master) %>%
  ph_with(value = ft_data, location = ph_location_type(type = "body"))
```

Although this solution works fine, it does not allow formatting the table as much as we would want. Instead, we prefer to use another package called {`flextable`} (see Section 5.3.1 for an introduction) which was developed by the same author as {`officer`}.

Remember that with {`flextable`}, the procedure starts with creating a *flextable* object (here `ft_table`) using the `flextable()` function.

```
library(flextable)

ft_table <- ft_data %>%
  arrange(Product) %>%
  flextable()
```

This table can be customized in various ways such as:

- `align()` and `rotate()` controls for the text alignment and its rotation;
- `bold()` and `italic()` writes the text in bold and italic;
- `font()` and `fontsize()` controls the font type and the size to use;
- `color()` and `bg()` allows changing the color of the text and the background.

All these functions require informing the rows (parameter `i`) and the columns (`j`) as well as the `part` (`"body"`, `"header"`, `"footer"`, or `"all"`) to modify.

---

[7] `senso_mean` was built in Section 6.2.1.

Additionally, further formatting can be applied to the table itself through the following functions:

- `height()` & `width()` control the row height and column width;
- `border_outer()`, `border_inner()`, `border_inner_h()` and `border_inner_v()` help design the table by adding borders;
- `autofit()` and `padding()` are used to control the final size of the table.

For illustration, let's apply some of these functions to `ft_table`:

```
ft_table <- ft_table %>%
  fontsize(size = 11) %>%
  # Formatting the header
  font(fontname = "Roboto", part = "header") %>%
  color(color = "white", part = "header") %>%
  bold(part = "header") %>%
  align(align = "center", part = "header") %>%
  bg(bg = "#324C63", part = "header") %>%
  # Formatting the body
  font(fontname = "Calibri", part = "body") %>%
  bg(i = 1:nrow(ft_data), bg = "#EDEDED") %>%
  # Formatting the last row of the table
  bold(i = nrow(ft_data), j = 1:ncol(ft_data)) %>%
  italic(i = nrow(ft_data), j = ~Product+Salty+Sweet+Sour+Bitter) %>%
  color(i = nrow(ft_data), j = ~Sour, color = "red") %>%
  color(i = nrow(ft_data), j = ~Sweet, color = "orange") %>%
  autofit()

# Set up the border style
my_border <- fp_border(color = "black", style = "solid", width = 1)

ft_table <- ft_table %>%
  border_outer(part = "all", border = my_border) %>%
  border_inner(part = "body", border = fp_border(style = "dashed")) %>%
  width(j = 1, width = 1.2)
```

This is just an overview of the most relevant and used functions in {`flextable`}, yet there are more possibilities. To go further, you can also consider the following functions (among many more):

- `merge()` merges vertically or horizontally cells with the same content;
- `compose()`, `as_chunk()`, and `as_paragraph()` work hand in hand to create more complex text formatting (e.g. sentence with parts of the text colored differently or with sub/superscript);
- `style()` applies a set of formatting properties to the same selection of the rows/columns.

Finally, to export a *flextable* table to a PowerPoint deck, simply export it as we have seen before:

```
pptx_obj <- pptx_obj %>%
  add_slide(layout = "Title and Content", master = master) %>%
  ph_with(value = ft_table,
          ph_location(left = 2, top = 2, width = 4))
```

### Exporting Plots

The last type of R outputs to export to PowerPoint are figures. Before showing how to export them, let's build a simple bar chart from `senso_mean` using {`ggplot2`}:

```
chart_to_export <- senso_mean %>%
  dplyr::select(Product, Salty, Sweet, Sour, Bitter) %>%
  arrange(Product) %>%
  pivot_longer(Salty:Bitter,
               names_to = 'Attribute', values_to = 'Value') %>%
  ggplot(aes(x = Product, y = Value, fill = Attribute)) +
  geom_col(position = 'dodge')+
  xlab("")+
  theme_bw()
```

To export any *ggplot2* object to PowerPoint, the package {`rvg`} is required. This package provides two graphic devices that produce Vector Graphics outputs in *DrawingML* format for Microsoft PowerPoint with `dml_pptx()` and for Microsoft Excel with `dml_xlsx()`, meaning the graphics is being "rebuilt" in PowerPoint or Word. To simplify, the generic `dml()` function is used, and depending on the output format, the corresponding function is being called.

```
library(rvg)
pptx_obj <- pptx_obj %>%
  add_slide(layout = "Title and Content", master = master) %>%
  ph_with(value = dml(ggobj = chart_to_export),
          location = ph_location_type(type = 'body'))
```

With {`rvg`}, the graphics are being *rebuilt* in PowerPoint, meaning that they are completely editable. It is hence possible to change color, rewrite text, move labels, etc.

To go further, the {mschart} package creates the graphs directly in PowerPoint or Word. These graphics have then the advantage to be interactive. However, this package is only limited to simple graphics (such as line chart, bar charts, etc.)

To produce such interactive graphs, `ggplot2` graphs are not needed. Instead, functions such as `ms_barchart()` are called to produce them.

```
library(mschart)

mydata <- senso_mean %>%
  dplyr::select(Product, Salty, Sweet, Sour, Bitter) %>%
  arrange(Product) %>%
  pivot_longer(Salty:Bitter,
               names_to = 'Attribute', values_to = 'Value')

# Building the barchart using ms_barchart()
my_barchart <- ms_barchart(data=mydata, x="Product", y="Value",
                           group="Attribute")

# The chart is a PowerPoint native object
# It can be viewed using the preview option in print
print(my_barchart, preview = TRUE)
```

```
## * "ms_barchart" object
##
## * original data [44,3] (sample):
##   Product Attribute  Value
## 1     P01     Salty  5.100
## 2     P01     Sweet 22.200
## 3     P01      Sour  0.000
## 4     P01    Bitter  8.000
## 5     P02     Salty  2.933
##
## * series data [11,5] (sample):
##   Product Bitter Salty Sour Sweet
## 1     P01  8.000 5.100    0  22.2
## 2     P02  4.933 2.933    0  15.8
## 3     P03  7.800 4.667    0  10.4
## 4     P04  4.267 3.600    0  16.6
## 5     P05  6.733 5.867    3  21.0
```

```
# To add the object to a PPT slide, officer's ph_with() function is used
pptx_obj <- pptx_obj %>%
  add_slide(layout = "Title and Content", master = "Office Theme") %>%
  ph_with(value = my_barchart,
          location = ph_location_type(type = "body"))
```

Now the full deck is being created, let's save it one last time using `print()`:

```
print(pptx_obj, target = "output/my export.pptx")
```

If you open the PowerPoint just exported, on the final slide, you'll find the bar chart generated by {mschart}. By clicking the graph, you'll find a "funnel" icon on the right side, which allows you filter attributes or products, hence making your graph interactive.

At last, {officer} also allows you adding images that are stored on your computer into a PowerPoint deck. This can be done through the `external_img()` function, which takes as input the location of the file. Like for any other graph, simply apply this function within `ph_with()` by specifying the location where to print the image.

### 6.2.3 Word

The process for building Word document directly from R is very similar to the one for PowerPoint, since it is also handled though {officer}.

To start a new Word document, the `read_docx()` function is being used. Since Word documents are more *text oriented* than PowerPoint, blocks of text are defined as paragraph. To introduce a new paragraph, the `body_add_par()` function is called. Note that paragraphs are automatically separated by line breaks:

```
docx_obj <- read_docx() %>%
  body_add_par(value = "My Text", style = "Normal") %>%
  body_add_par(value = "Other Text", style = "Normal") %>%
  body_add_par(value = "Conclusion", style = "Normal")
```

Here again, the results can be exported to your computer using `print()`:

```
print(docx_obj, target = "output/my export.docx")
```

Of course, it is not required to use the default formatting options from the word document in use. Instead, we can format it directly from R using `body_add_fpar()` to add a formatted text paragraph, or apply predefined styles to the previous function suggested (as is the case here with `style = "heading 1"` to set the text as a title of level 1).

```
my_format <- fp_text(font.family = 'Calibri', font.size = 14,
                     bold = TRUE, color = 'blue')
my_text <- ftext('Here is another example of text', my_format)
my_par <- fpar(my_text)

docx_obj <- read_docx() %>%
  body_add_par(value = "Document Title", style = "heading 1") %>%
  body_add_par(value = "", style = "Normal") %>%
  body_add_fpar(my_par, style = "Normal")
```

To export tables or figures, additional functions including `body_add_table()` (for tables) and `body_add_gg()`[8] (for `ggplot()` figures) are used. These can be combined to `body_add_caption()` to add a caption to your table/figure:

```
table_num <- run_autonum(seq_id = "tab",
                         pre_label = "Table ", bkm = "tables")
figure_num <- run_autonum(seq_id = "fig",
                          pre_label = "Figure ", bkm = "figures")

docx_obj <- docx_obj %>%
  body_add_par(value = "Exporting Tables",
               style = "heading 2") %>%
  body_add_par(value = "",
               style = "Normal") %>%
  body_add_par(value = "Here is my first table:",
               style = "Normal") %>%
  body_add_par(value = "",
               style = "Normal") %>%
  body_add_table(value = head(mtcars)[,1:4],
                 style = "table_template") %>%
  body_add_caption(block_caption("My first table.",
                                 style="centered",
                                 autonum=table_num)) %>%
  body_add_par(value = "Exporting Figures",
               style = "heading 2") %>%
  body_add_par(value = "",
               style = "Normal") %>%
  body_add_par(value = "Here is my first figure:",
               style = "Normal") %>%
  body_add_par(value = "",
               style = "Normal") %>%
  body_add_gg(value = chart_to_export) %>%
  body_add_caption(block_caption("My first figure.",
                                 style="centered",
                                 autonum=figure_num))
```

[8] Note that `body_add_img()` and `body_add_plot()` can also be used.

As can be seen, `body_add_caption()` is combined to `block_caption()` and have some automated numbering, as defined previously using `table_num` for tables and `figure_num` for figures.

Unlike a PowerPoint file that contains separate slides, a word document is a continuous object. Hence, to emphasize a break and add content to a new page, `body_add_break()` needs to be called. Additionally, tables of content can be generated using `body_add_toc()`:

```
docx_obj <- docx_obj %>%
  body_add_break() %>%
  body_add_par(value = "Conclusion", style = "heading 1") %>%
  body_add_break() %>%
  body_add_par("Table of Contents", style = "heading 1") %>%
  body_add_toc(level = 2)
```

Finally, let's export this last version of the word document to visualize it:

```
print(docx_obj, target = "output/my export.docx")
```

As can be seen, it is possible to format a nice report in Word directly from R, that integrates text, tables, figures, and more. For more information regarding {`officer`} and on how to export results to Word and PowerPoint, please visit https://ardata-fr.github.io/officeverse/index.html.

---

It is worth mentioning that {`officer`} also allows extracting information from existing reports (Word and PowerPoint). It is, however, outside the scope of this book and will not be further described.

---

### 6.2.4 Notes on Applying Corporate Branding

You may have noticed that we have been consistent with our approach to export results to reports, regardless of the final output: We start with predefining our styling parameters that we then apply to our different tables, slides, paragraphs, etc. This is not a formal rule, yet we strongly recommend you adopting this way of working. Indeed, by creating your different styling parameters at the start of your script file, these lines of code do not interfere with your analyses. At a later stage, you will thank yourself for keeping a well-structured code as it gains in clarity, and hence facilitates debugging your code in case of error or changes.

To go one step further, we would recommend you storing all these styling parameters in a separate file you load any time you need them through `source()`. This process reduces the size of your script file, hence increasing its clarity, while harmonizing all your exports by centralizing your formatting code in one unique place. The last point is crucial since any changes only need to be done once, and yet will be applied to all your reports.

As we have seen, {`officer`} gives you the opportunity to import predefined templates (PowerPoint or Word). This is very valuable as your report can easily match your corporate style.

Ultimately, to ensure optimal efficiency, we advise you to spend a bit more time when constructing your report by ensuring that as many details are being taken care of, so that later on, you can spend more time in the story building part and less on the analysis and slide generation. For instance, don't be afraid of mass-exporting results, as it is easier to remove slides, tables, or figures (in case they are not needed for your story) than it is to regenerate them at a later stage (if missing).

## 6.3 Integrating Analyses Scripts Within Your Reporting Tool

As we have just seen, we can generate reports in the Microsoft Office Suite directly from our R script. Although the results are being showed, the script used to reach these results is completely hidden. Of course, we could add them as text, but the logic would suggest that the researcher can just get back to the script to decode how certain outputs have been obtained.

Let's now change our way of thinking by proposing an alternative in which we integrate our R analysis directly within a reporting tool. For that, we need to introduce another useful package for reporting and document building: {`rmarkdown`}.

### 6.3.1 What Is {rmarkdown}

Markdown is an ecosystem specific to text document, in which authors *script* their reports by controlling various features including:

- paragraphs and inline formatting (e.g. bold, italic, etc.);
- (section) headers;
- blocks (code or quotations);
- (un)numbered lists;
- horizontal rules;
- tables and figures (including legends);

- LaTeX math expressions, formulas, and theorems;
- links, citations, and footnotes.

Limiting the creation of a Markdown document to this list of elements is more an advantage than a drawback as it suffices to create technical and non-technical documents while still keeping it simple.

In practice, R Markdown provides an authoring framework for data science, as it can be used for saving/documenting/executing code and generating high-quality reports. Once the document is being created, you can then compile it to build it in the output format of your choice (e.g. word, pdf, html, etc.)

### 6.3.2 Starting with `{rmarkdown}`

To start, you need to install the `{rmarkdown}` package using the `install.packages()` function. To load this package, just type `library(rmarkdown)`. If you intend to build your report in pdf, you also need to install a LaTeX library. For its simplicity, we recommend you installing the TinyTeX library using `install.packages("tinytex")`.

Let's start with a simple example that is provided by RStudio. To start a RMarkdown document, click *File > New File > R Markdown...* This opens a new window in which you can inform the name of your file, the author name, and the type of report to create (HTML, PDF, or Word). Once set, click *OK*. A new script file of type *.Rmd* opens.

In this document, there are three components: metadata, text, and code.

The document starts with the metadata. It is easily recognizable as it starts and ends with three successive dashes (`---`), and its syntax is YAML (YAML Ain't Markup Language). In this part, information regarding the properties of the final document is being stored. This includes (among other) the title, authorship, date, export format, etc. of the final document.

Be aware that indentation matters in YAML, so follow the rules to ensure that your document compiles correctly.

Right after the metadata is the body of document. The syntax for the text is Markdown, and the main features will be presented in Section 6.3.3. Within the body, computer code can be added, either as a chunk, or within the text.

### 6.3.3 `{rmarkdown}` through a Simple Example

To illustrate the use of `{rmarkdown}`, let's consider this simple document (inspired from Yihuie Xie, J. J. Allaire, & Garrett Grolemund (2019). *R Markdown. The Definitive Guide.* CRC Press) entitled RMarkdown example.Rmd and stored in the directory of this book.

The top of the document contains the metadata, which (in our case) will generate the report in an HTML document.

Next, we have a first chunk of code that sets the main options on how the code should be handled. If all the code chunks are handled in the same way, it is handy to set it at the start. However, when different chunks of code should be handled differently, it may be easier to define for each section how it should be handled.

There are mainly four ways to handle code.

> The first way is defined here on the code chunk *header* as `include = FALSE`[9]: `include` always run the code, yet it allows printing (`include = TRUE`) or not (`include = FALSE`) the code and its outputs in the final document.
>
> The second option is `echo`. In this code chunk, we automatically set that all the code chunk should be defined as `echo = TRUE`, which means that the code will run and be printed (together with its output) in the document. This seems very similar to `include`, yet it differs from it as `echo = FALSE` runs the code, prints the outputs, but not the code.
>
> If you only want to show some code without running it, the `eval` parameter is used (`eval = FALSE` means that the code will be displayed but will not run). This is useful for displaying example code or for disabling large or time-consuming chunk of codes without having to set it up as comment.
>
> Last, we can control whether outputs should be shown or hidden using `results` (printed output) and `fig.show` (plots). By default, the results are shown, unless it is set as `results = "hide"` or `fig.show = "hide"`.

The document then continues with a section header, which starts with #. The hierarchy of headers is defined by the number of adjacent # (for a header of level 3, starts the header with ###).

In this section, a first paragraph is being written. This is plain text, except for two particular words, one written between two "`" (backticks) and one written between two double "*" (stars). Here, the backticks are used to write text in R font (or as we will see later, to render results from R), whereas the double stars write the text in bold (double "_" (underscore) could also be used). For italic, one single star (or one single underscore) is used.

For the following section and subsections, we introduce numbered and unnumbered list of elements. For numbered list, it starts with a *number* followed by a . (numbers will be incremented automatically). For unnumbered lists, you can either start with a "-" (dash), or with "*" (star) for bullet points. For sub-lists, indent your marker by pressing the *Tab* key.

In the next section called *Analysis*, we are running our first lines of code.

---

[9] When set manually, this is where you should indicate how to handle each chunk of code.

The first code chunk runs a regression model. In the text under the second chunk of code, we are retrieving automatically a value from R by including a *r* at the starts of two *backticks* followed by the element to retrieve. In our final report, this code will automatically be replaced by the value 3.93.

The second code chunk shows how the results can be printed, either directly from R or in a nicer way using the `knitr::kable()` function.

Finally, the last code chunk of this section creates a plot with a caption that is automatically numbered.

### 6.3.4 Creating a Document Using {`knitr`}

Once the document is ready, you can neat it using the `knit` button. This will create the report in the format of interest (here HTML).

### 6.3.5 Example of Applications

{`rmarkdown`} is a very powerful tool for building report, in particular in the context of reproducible research since it allows sharing code and running analyses within the report (part of the text around the code can justify the decisions made in terms of analyses to ensure transparency). The latter point is particularly interesting since any change in the data will automatically provide updated results throughout the report, without you having to change them manually.

Its applications are various and can go from report to teaching material, publication, or even books (this book has been written in {`rmarkdown`} and its extension {`bookdown`}), emails, websites, dashboards, surveys, etc. Even more interestingly, {`rmarkdown`} can also be combined to {`shiny`} to build interactive reports, dashboards, or teaching materials in which users would (say) import their data set, select the variables to analyze through buttons, chose which analyses and which options to perform, and the results will automatically be generated accordingly.

For more information on {`rmarkdown`} and related packages, please see:

- Bookdown https://bookdown.org/yihui/bookdown/
- R Markdown https://bookdown.org/yihui/rmarkdown/
- R Markdown Cookbook https://bookdown.org/yihui/rmarkdown-cookbook/

As mentioned earlier, R Markdown can also be used to generate other types of documents, including presentations. This can be done directly from the {`rmarkdown`} package using *ioslides* presentation (`output: ioslides_presentation` in the metadata), *Slidy* presentation (`output: slidy_presentation`), or *PowerPoint* presentation (`output: powerpoint_presentation` with `reference_doc: my-styles.pptx` to apply your own template) just to name a few. It can also be done using additional packages such as {`xarigan`}.

## 6.4 To Go Further...

If R allows you saving time by creating your report within your R-script or by running your analysis within your report document, it cannot communicate the presentation to your partners/clients for you. However, if the report is very standard (say only key results, tables or figures) or running routinely (say in quality control), R could automatically run the analysis as soon as new data is available, build the report, and send it automatically to you, your manager, or your colleagues and partners by email.

Such process can be done thanks to the {`blastula`} package.

# 7

# *Example Project: The Biscuit Study*

## 7.1 Objective of the Test

The data set that we use as a main example throughout this book comes from a sensory study on biscuits. The study was part of the project BISENS funded by the French National Research Agency (ANR, programme ALIA 2008). These biscuits were developed for breakfast consumption and specifically designed to improve satiety.

The study was conducted in France with 107 consumers who tested a total of 10 biscuit recipes (including 9 experimental products varying in their fiber and protein contents, as fibers and proteins are known to increase satiety).

The study aimed to measure the liking for these biscuits, its link with eaten quantities, and the evolution of hunger sensations over ad libitum consumption. All the volunteers therefore participated to 10 morning sessions in order to test every product (one biscuit type per session). After they completed all the sessions, they also filled a questionnaire about food-related personality traits such as cognitive restraint and sensitivity to hunger.

Parallel to this, a panel of nine trained judges performed a quantitative descriptive analysis of the biscuits. They evaluated the same 10 products as well as an additional product whose recipe was optimized for liking and satiating properties.

Data from the biscuit study are gathered in three Excel files that can be accessed here https://github.com/aigorahub/data_science_for_sensory:

- biscuits_consumer_test.xlsx
- biscuits_sensory_profile.xlsx
- biscuits_traits.xlsx

DOI: 10.1201/9781003028611-7

## 7.2 Products

In total, 11 products were considered in this study. They are all breakfast biscuits with varying contents of proteins and fibers (Table 7.1). Products *P01* to *P09* are prototypes whereas product *P10* is a standard commercial biscuit without enrichment. The 11th product *Popt* is an additional optimized biscuit that has been evaluated only by the trained panel for descriptive analysis.

**TABLE 7.1**

Product set for the biscuit study

| Product | Protein | Fiber | Type |
|---|---|---|---|
| P01 | Low | Low | Trial |
| P02 | Low | High | Trial |
| P03 | High | High | Trial |
| P04 | High | High | Trial |
| P05 | High | Low | Trial |
| P06 | High | Low | Trial |
| P07 | Low | High | Trial |
| P08 | High | Low | Trial |
| P09 | High | High | Trial |
| P10 | Low | Low | Commercial product |
| POpt | High | Low | Optimized trial |

## 7.3 Sensory Descriptive Analysis

A panel of 9 trained judges evaluated the 11 products on 32 sensory attributes (8 attributes for aspect, 3 for odor, 12 for flavor, and 9 for texture).

Judges had received thorough training for the wider biscuit product category over more than six months on common vocabulary following conventional descriptive analysis standards (ISO11035, 1995; ISO8586, 2012). Evaluation was performed in sensory booths in triplicates, although only individual means are available in this data set.

For each product, the judges reported the perceived intensity of each attribute using unstructured linear scales on a computer screen. Intensities were automatically converted by the acquisition system into a score ranging from 0 to 60. These data are stored in *biscuits_sensory_profile.xlsx*.

In summary, the sensory data have 99 observations (11 *Products* × 9 *Judges*) and 34 variables (*Judge* and *Product* identifiers + 32 attributes).

## 7.4 Consumer Test

### 7.4.1 Participants

107 women who were all regular consumers of breakfast biscuits participated in the test. The *biscuits_traits.xlsx* file gives information about their Body Mass Index (BMI) (Q4–Q6) and their socio-demographics (Q7–Q11: marital status, household, income, occupation, and highest degree).

This file also gives participants' answers to a self-assessment questionnaire (Q12–62) that evaluates eating behavioral traits with emphasis on the tendency to control food intake cognitively. The questionnaire comprises a series of assertions about various eating situations in the respondent's daily life (e.g. "How...." answer Y/N"). Resulting scores are loaded into three factors: cognitive restraint (conscious restriction of food intake in order to control body weight or to promote weight loss), disinhibition (or emotional eating), and susceptibility to hunger (or uncontrolled eating, i.e. tendency to eat more than usual due to a loss of control over intake accompanied by feelings of hunger). This questionnaire is thus known as the Three-Factor Eating Questionnaire (TFEQ) (Stunkard and Messick, 1985) and is one of the most commonly used questionnaires to evaluate eating behaviors in relation to overweight or obesity (Blundell et al., 2010). Calculation of these factors is detailed in Chapter 10.

---

In summary, the data stored in *biscuits_traits.xlsx* have 107 observations and 62 variables giving multiple information about the consumers, including their sociodemographics, BMI, and TFEQ answers. Variable codes, names, and meaning are listed in the second sheet "Variables", with their corresponding levels in the third sheet "Levels".

---

### 7.4.2 Test Design

The presentation order of the different products was randomized across the panel. Again, consumers evaluated one biscuit type per day/session.

The design of the sessions is summarized in Figure 7.1 with main measured variables. After they first rated their appetite sensations using visual analog scales (VAS), the participants tasted and rated one biscuit for liking. They were then served with a box of the same biscuits for *ad libitum* consumption (with a maximum of 10 biscuits), followed by a new questionnaire regarding their liking, pleasure, and appetite sensations.

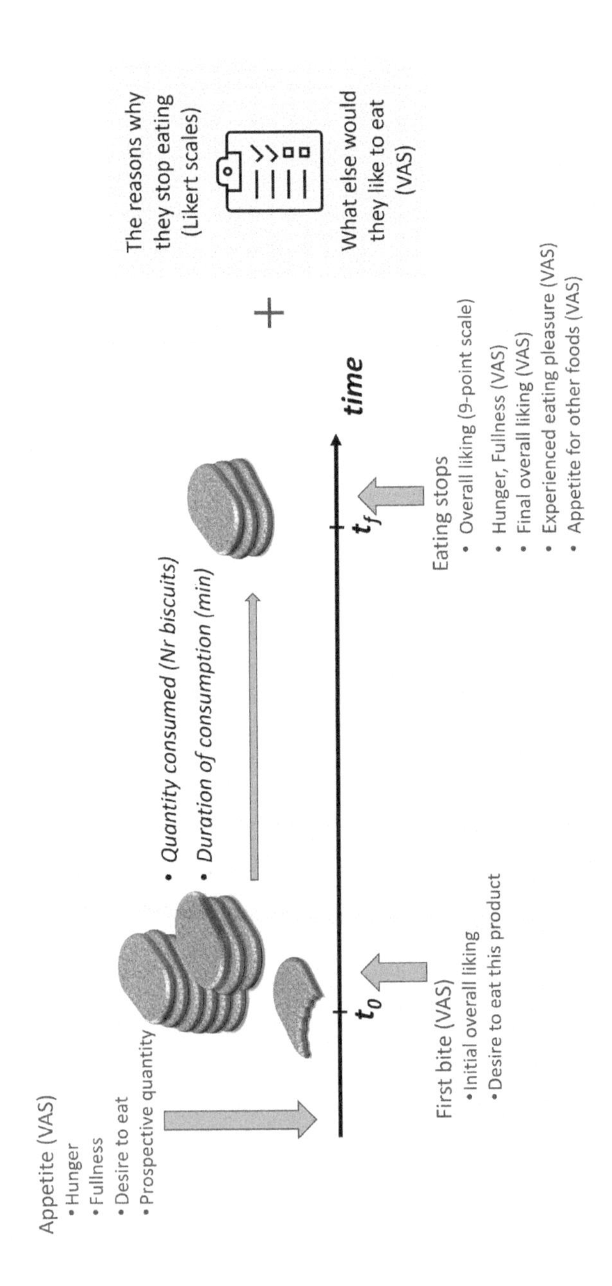


**FIGURE 7.1**
General design for the consumer test of the biscuit study. Participants were served with a different set of biscuits every session.

### 7.4.3 Evaluation

The liking was measured with two different scales:

1. with a horizontally oriented unstructured linear scale (i.e. VAS) anchored with *"I don't like this biscuit at all"* (left end) and *"I like this biscuit a lot"* (right end) at two different times: after the first bite and at the end of their consumption.
2. with a vertically oriented semantic nine-point hedonic scale when stopping their consumption.

VAS scales are frequently used in nutrition studies (Stubbs et al., 2000), whereas the nine-point hedonic scale is more popular in sensory and consumer science (Peryam and Pilgrim, 1957; Wichchukit and O'Mahony, 2015).

Once done, the participants were asked about the reason(s) why they stopped eating (6 potential reasons rated with *Likert* scales ranging from *strongly disagree* to *strongly agree*). They were also asked how much they would like to eat other types of foods (11 food items rated using a VAS).

The time spent in the booth and the number of biscuits eaten by each participant were recorded by the experimenters, as well as the type of drink they selected and the approximate volume they drank during each session. These data are stored in *biscuits_consumer_test.xlsx*, in the second tab named *Time Consumption*.

---

In summary, the consumer test data have 1070 observations (10 *Products* × 107 *Consumers*) and 32 variables (including *Consumer* and *Product* identifiers). The *biscuits_consumer_test.xlsx* file also includes one sheet with consumption time and the number of biscuits that consumers ate, and a sheet indicating the average weight for each product.

---

# 8

# *Data Collection*

Before any statistical analysis and vizualizations, robust data need to be collected. This important step often requires a *proper* experimental design, that is, an experimental design that would assure relevant and meaningful data are obtained with maximum efficiency to answer our research questions. This chapter approaches all the required steps to reach such goal, from setting up the test (e.g. estimation of the number of panelists, design of sensory evaluation sessions and design of experiments), to the collection of data (through valuable execution tips) and its importation in a statistical software (R, here).

## 8.1 Designs of Sensory Experiments

Like with any other chapter, let's start by loading the {tidyverse}.

```
library(tidyverse)
```

### 8.1.1 General Approach

Sensory and consumer science relies on experiments during which subjects usually evaluate several samples one after the other. This type of procedure is called "monadic sequential" and is common practice for all three main categories of tests (difference testing, descriptive analysis, and hedonic testing). The main advantage of proceeding this way is that responses are within-subjects (data can be analyzed at the individual level) so that analysis and interpretation can account for inter-individual differences, which is a constant feature of sensory data.

DOI: 10.1201/9781003028611-8

However, this type of approach also comes with drawbacks[1] as it may imply order effects and carry-over effects (Macfie et al., 1989). Fortunately, most of these effects can be controlled with a proper design of experiment (DoE). A good design ensures that order and carry-over effects are not confounded with what you are actually interested to measure (most frequently, the differences between products) by balancing these effects across the panel. However, it is important to note that the design does not eliminate these effects and that each subject in your panel may still experience an order and a carry-over effect, as well as boredom, sensory fatigue, etc.

Before going any further into the design of sensory evaluation sessions, it is important to first estimate the number of panelists needed for your study. For that, you may rely on common practices. For instance, most descriptive analysis studies with trained panelists are typically conducted with 10–20 judges, whereas 100 participants is usually considered as a minimum for hedonic tests. Of course, these are only ballpark numbers and they must be adjusted to the characteristics of the population you are interested in and to the specifics of your study objectives. In all cases, a power analysis would be wise to make sure that you have a good rationale for your proposed sample size, especially for studies involving consumers. The {pwr} package provides a very easy way to do that, as shown in the example code below for a comparison between two products on a paired basis (such as in monadic sequential design). Note that you need to provide an effect size (expressed here by Cohen's *d*, which is the difference you aim to detect divided by the estimated standard deviation of your population).

```
library(pwr)
pwr.t.test(n=NULL, sig.level=0.05, type="paired",
           alternative="two.sided", power=0.8, d=0.3)
```

```
##
##      Paired t test power calculation
##
##               n = 89.15
##               d = 0.3
##       sig.level = 0.05
##           power = 0.8
##     alternative = two.sided
##
## NOTE: n is number of *pairs*
```

[1] Market researchers would argue that evaluating several products in a row doesn't usually happen in real life and that proceeding this way may induce response biases. They thus advocate the use of pure monadic designs in which participants are only given one sample to evaluate. This corresponds to a between-group design that is also frequently used in fields where only one treatment per subject is possible (drug testing, nutrition studies, etc.).

For discrimination tests (e.g. tetrad, 2-AFC, etc.), the reader may also refer to the {sensR} package and its `discrimSS()` function for the sample size calculation in both difference or similarity testing contexts.

### 8.1.2 Crossover Designs

For any sensory experiment that implies the evaluation of more than one sample, first-order and/or carry-over effects should be expected. That is to say, the evaluation of a sample may affect the evaluation of the next sample even though sensory scientists try to lower such effects by asking panelists to pause between samples and use of appropriate mouth-cleansing techniques (drinking water, eating unsalted crackers, or a piece of apple, etc.). The use of crossover designs is thus highly recommended (Macfie et al., 1989).

Williams's Latin-Square designs offer a perfect solution to balance carry-over effects. They are very simple to create using the `williams()` function from the {crossdes} package. For instance, if you have five samples to test, `williams(5)` would create a 10x5 matrix containing the position at which each of three samples should be evaluated by 10 judges (the required number of judges per design block).

Alternately, the `WilliamsDesign()` function in {SensoMineR} allows you to create a matrix of samples (as numbers) with numbered `Judges` as row names and numbered `Ranks` as column names. You only have to specify the number of samples to be evaluated, as in the example below for five samples.

```
library(SensoMineR)
wdes_5P10J <- WilliamsDesign(5)
```

```
##          Rank 1 Rank 2 Rank 3 Rank 4 Rank 5
## Judge 1       3      2      5      4      1
## Judge 2       2      4      3      1      5
## Judge 3       4      1      2      5      3
## Judge 4       1      5      4      3      2
## Judge 5       5      3      1      2      4
## Judge 6       1      4      5      2      3
## Judge 7       5      1      3      4      2
## Judge 8       3      5      2      1      4
## Judge 9       2      3      4      5      1
## Judge 10      4      2      1      3      5
```

Suppose you want to include 20 judges in the experiment, you would then need to duplicate the initial design.

```
wdes_5P20J <- do.call(rbind, replicate(2, wdes_5P10J, simplify=FALSE))
rownames(wdes_5P20J) <- paste("judge", 1:20, sep="")
```

```
##         Rank 1 Rank 2 Rank 3 Rank 4 Rank 5
## judge1       3      2      5      4      1
## judge2       2      4      3      1      5
## judge3       4      1      2      5      3
## judge4       1      5      4      3      2
## judge5       5      3      1      2      4
## judge6       1      4      5      2      3
## judge7       5      1      3      4      2
## judge8       3      5      2      1      4
## judge9       2      3      4      5      1
## judge10      4      2      1      3      5
## judge11      3      2      5      4      1
## judge12      2      4      3      1      5
## judge13      4      1      2      5      3
## judge14      1      5      4      3      2
## judge15      5      3      1      2      4
## judge16      1      4      5      2      3
## judge17      5      1      3      4      2
## judge18      3      5      2      1      4
## judge19      2      3      4      5      1
## judge20      4      2      1      3      5
```

The downside of Williams's Latin square designs is that the number of samples ($k$) to be evaluated dictates the number of judges. For an even number of samples you must have a multiple of $k$ judges and a multiple of $2k$ judges for an odd number of samples.

As the total number of judges in your study may not always be exactly known in advance (e.g. participants not showing up to your test, extra participants recruited at the last minute), it can be useful to add some flexibility to the design. Of course, additional rows would depart from the perfectly balanced design, but it is possible to optimize them using Federov's algorithm thanks to the `optFederov()` function of the {`AlgDesign`} package, by specifying `augment = TRUE`. For example we can add three more judges to the Williams Latin square design that we just built for $nbP = 5$ products and 10 judges, hence leading to a total number of $nbP = 13$ judges. Note that this experiment is designed so that each judge will evaluate all the products; therefore, the number of samples per judge ($nbR$) equals the number of products ($nbP$).

```
library(AlgDesign)
nbJ=13
nbP=5
nbR=nbP

wdes_5P10J <- WilliamsDesign(nbP)
tab <- cbind(prod=as.vector(t(wdes_5P10J)),
```

```
                judge=rep(1:nbJ,each=nbR),
                rank=rep(1:nbR,nbJ))
optdes_5P13J <- optFederov(~prod+judge+rank, data=tab, augment=TRUE,
                           nTrials=nbJ*nbP, rows=1:(nbJ*nbP),
                           nRepeats = 100)
xtabs(optdes_5P13J$design)
```

```
##      rank
## judge 1 2 3 4 5
##    1  1 3 4 2 5
##    2  3 2 1 5 4
##    3  2 5 3 4 1
##    4  5 4 2 1 3
##    5  4 1 5 3 2
##    6  5 2 4 3 1
##    7  4 5 1 2 3
##    8  1 4 3 5 2
##    9  3 1 2 4 5
##    10 2 3 5 1 4
##    11 1 3 4 2 5
##    12 3 2 1 5 4
##    13 2 5 3 4 1
```

In the code above, `xtabs()` is used to arrange the design in a table format that is convenient for the experimenter.

Note that it would also be possible to start from an optimal design and expand it to add one judge at a time. The code below first builds a design for 5 products and 13 judges and then adds one judge to make the design optimal for 5 products and 14 judges.

```
nbJ=13
nbP=5
nbR=nbP

optdes_5P13J <- optimaldesign(nbP, nbP, nbR)$design
tab <- cbind(prod=as.vector(t(optdes_5P13J)),
             judge=rep(1:nbJ,each=nbR),
             rank=rep(1:nbR,nbJ))
add <- cbind(prod=rep(1:nbP,nbR),
             judge=rep(nbJ+1,nbP*nbR),
             rank=rep(1:nbR,each=nbP))

optdes_5P14J <- optFederov(~prod+judge+rank, data=rbind(tab,add),
                           augment=TRUE, nTrials=(nbJ+1)*nbP,
                           rows=1:(nbJ*nbP), nRepeats = 100)
```

### 8.1.3 Balanced Incomplete Block Designs (BIBD)

Sensory and consumer scientists may sometimes consider using incomplete designs, that is, experiments in which each judge evaluates only a subset of the complete product set (Wakeling and MacFie, 1995). In this case, the number of samples evaluated by each judge remains constant but is lower than the total number of products included in the study.

You might want to choose this approach, for example, if you want to reduce the workload for each panelist and limit sensory fatigue, boredom, and inattention. It might also be useful when you cannot "afford" a complete design because of sample-related constraints (limited production capacity, very expensive samples, etc.). The challenge then is to balance sample evaluation across the panel as well as the context (i.e. other samples) in which each sample is being evaluated. For such a design, you thus want each pair of products to be evaluated together the same number of times.

The `optimaldesign()` function of {`SensoMineR`} can be used to search for a Balanced Incomplete Block Design (BIBD). For instance, let's imagine that 10 panelists are evaluating 3 out of 5 possible samples. The design can be defined as follows:

```
incompDesign1 <- optimaldesign(nbPanelist=10,
                               nbProd=5,
                               nbProdByPanelist=3)
incompDesign1$design
```

```
##             Rank 1 Rank 2 Rank 3
## Panelist 1       4      1      5
## Panelist 2       3      2      5
## Panelist 3       3      5      4
## Panelist 4       5      4      2
## Panelist 5       5      3      1
## Panelist 6       1      2      3
## Panelist 7       1      4      2
## Panelist 8       4      1      3
## Panelist 9       2      3      4
## Panelist 10      2      5      1
```

BIBD are only possible for certain combinations of number of treatment (products), number of blocks (judges), and block size (number of samples per judge). Note that `optimaldesign()` will yield a design even if it is not balanced but it will also generate contingency tables allowing you to evaluate the design's orthogonality and how well balanced are order and carry-over effects.

You can also use the {crossdes} package to generate a BIBD with this simple syntax: find.BIB(trt, b, k, iter), with trt the number of products (here 5), b the number of judges (here 10), k the number of samples per judge (here 3), and iter the number of iteration (30 by default). Furthermore, the isGYD() function evaluates whether the incomplete design generated is balanced or not. If the design is a BIBD, you may then use williams.BIB() to combine it with a Williams design to balance carry-over effects.

```
library(crossdes)
incompDesign2 <- find.BIB(trt=5, b=10, k=3)
isGYD(incompDesign2)
williams.BIB(incompDesign2)
```

Incomplete balanced designs also have drawbacks. First, from a purely statistical perspective, they are conducive to fewer observations and thus to a lower statistical power. Product and Judge effects are also partially confounded even though the confusion is usually considered as acceptable.

### 8.1.4 Incomplete Designs and Sensory Informed Designs for Hedonic Tests

One may also be tempted to use incomplete balanced block designs for hedonic tests. However, proceeding this way is likely to induce framing bias. Indeed, each participant will only see part of the product set which would affect their frame of reference if the subset of the products they evaluate only covers a limited area of the sensory space.

Suppose you are designing a consumer test of chocolate chip cookies in which a majority of cookies are made with milk chocolate while a few cookies are made with dark chocolate chips. If a participant only evaluates samples that have milk chocolate chips, this participant will not know about the existence of dark chocolate and will potentially give very different scores compared to what they would have if they had a full view of the product category.

To reduce the risks incurred by the use of BIBD, an alternative strategy is to use a sensory informed design. Its principle is to allocate each panelist a subset of products that best cover the sensory diversity of the initial product set. Of course, this supposes that one has sensory data to rely on in the first place. Such data could derive from conventional descriptive analysis but rapid methods such as free sorting or napping could suffice. Practically, covering the sensory diversity amounts to maximizing the sensory distance between drawn products. Franczak et al. (2015) suggest an iterative algorithm to build such a design.

Ben Slama et al. (1998) have also proposed to select D-optimal subsets of products as a way to reduce the number of products in preference mapping studies.[2]

Selecting a D-optimal subset of products can be easily done with the Federov algorithm. Let's take, for example, the cocktail data set from the {SensoMineR} package. It contains a sensory descriptive analysis of 16 cocktails (`senso.cocktail`). Now, let's pretend that we want to conduct a hedonic test on these cocktails in the view of performing a preference mapping. Asking each consumer to evaluate all 16 cocktails may be considered to be too much (even if they don't contain alcohol!). We will thus use the `optFederov()` function of {AlgDesign} to select a D-optimal subset of, say, 8 cocktails. For that, we first need to run a PCA on the full sensory data. Then we use the coordinates of the products on the principal components (PC) to build our design. Here, we only pick the first two PC dimensions, but the analyst could select other dimensions that are deemed relevant to describe the product set (but keep in mind that increasing the number of dimensions will require a higher minimum number of products to be able to fit the corresponding models).

```
library(SensoMineR)
data(cocktail)

cocktail_pca <- senso.cocktail %>%
  PCA(graph = FALSE)

Prod_coord <- cocktail_pca$ind$coord[,1:2]
```

Assuming a full quadratic model for the preference mapping (see section 10.4.4), we can now use `optFederov()` to select a D-optimal subset of the desired number of products (here `ntrials=8`). The resulting selection will maximize the determinant of the information matrix (X'X) corresponding to the quadratic model. However, we can also decide to specify products that we want to be kept in the selection. Here, let's impose the selection of products 7 and 13.

[2] See also Rivière et al. (2006), for an interesting comparison of this approach with an adaptive sequential consumer test for preference analysis using Kano's model of satisfaction.

```
library(AlgDesign)
Dopt_des <- optFederov(~quad(.), Prod_coord, nTrials=8,
                       criterion="D",
                       augment = TRUE,
                       rows = c(7,13))
```

The resulting selection is listed in `Dopt_des$rows` and can be highlighted on the PCA score plot using {`factoextra`} and the `col.ind =` argument (Figure 8.1).

```
library(factoextra)

Prod_coord <- Prod_coord %>%
    as_tibble() %>%
    rownames_to_column("Product") %>%
    mutate(DOpt_subset = ifelse(Product %in% Dopt_des$rows,
                                "selected", "others"))

fviz_pca_ind(cocktail_pca,
             geom.ind = c("point", "text"),
             pointsize = 3,
             labelsize = 4, repel = TRUE,
             col.ind = Prod_coord$DOpt_subset,
             palette = c("#949494", "#00AFBB"),
             legend.title = "Subset",
             show.legend.text = FALSE,
             mean.point = FALSE
)
```

If needed, the coordinates of the D-optimal selection `Dopt_des$design` can be exported as a tibble and displayed with `flextable` (Table 8.1):

```
library(flextable)

Subset_coord <- Dopt_des$design %>%
                as_tibble() %>%
                mutate("Product" = factor(Dopt_des$rows), .before=1)

Subset_table <- flextable(Subset_coord) %>%
                colformat_double(digits = 2)
```

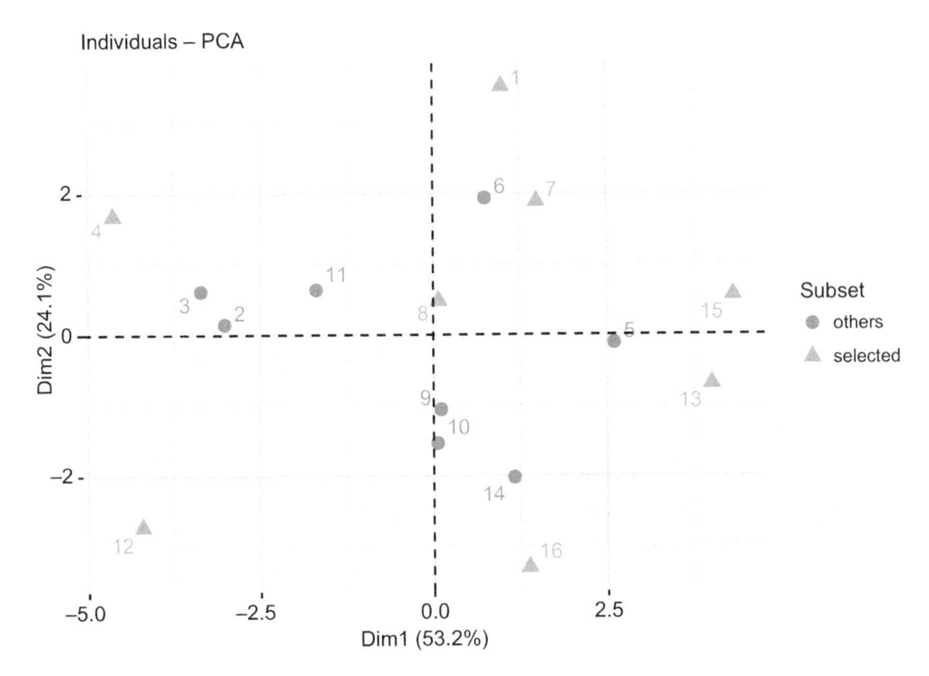


**FIGURE 8.1**
Highlighted D-optimal selection of products on the sensory map.

**TABLE 8.1**
Basic flextable

| Product | Dim.1 | Dim.2 |
|---|---|---|
| 1 | 0.99 | 3.51 |
| 4 | −4.63 | 1.68 |
| 7 | 1.49 | 1.89 |
| 8 | 0.08 | 0.48 |
| 12 | −4.21 | −2.72 |
| 13 | 3.99 | −0.70 |
| 15 | 4.31 | 0.55 |
| 16 | 1.38 | −3.30 |

Finally, it could be useful to calculate the determinant of the information matrix (assuming a quadratic model):

```
Xopti = model.matrix(~ quad(.), Dopt_des$design)
det(t(Xopti)%*%Xopti)
```

```
## [1] 3.414e+11
```

In itself, this value is not really telling. However, it can be compared to the determinant of the information matrix (X'X) for another subset (e.g., without the forced selection of two products, or a subset of different size) and see by how much it has improved or deteriorated. The determinant could also be used in an optimization loop to define complementary subsets to be allocated to different consumers or groups of consumers.

## 8.2 Product-related Designs

Because of their contribution to product development, sensory and consumer scientists often deal with DoE other than sensory designs strictly speaking (see for instance Gacula, 2008). Sensory-driven product development is indeed very frequent and implies strong interconnection between the measure of sensory responses and the manipulation of product variables (e.g. ingredients) or process variables (e.g. cooking parameters) (for a review, see Yu et al., 2018).

In order to get the most of sensory experiments, it is thus essential to ensure that the products or prototypes to be tested will be conducive to sound and conclusive data. First and foremost, as in any experimental science, one wants to avoid confounding effects. In addition to this and to put it more generally, the goal of DoE is to define which trials to run in order to be able to draw reliable conclusions without spending time and resources on unnecessary trials. In other words, one seeks maximum efficiency. This is especially critical in sensory science to limit the number of products to be evaluated and to keep panelists' workload under control.

### 8.2.1 Factorial Designs

Full factorial designs are of course commonly used and their application is usually straightforward. They won't be detailed here. However, it is worth noting that when the number of factors increases, the corresponding number of trials can quickly become daunting (e.g., $2^k$ trials for a two-level design with $k$ factors). Thus, always in the view of sparing experimental resources, incomplete, and fractional designs are frequently used.

Several strategies can be used to define which experiments to conduct (Dean et al., 2017; Lawson, 2014; Rasch et al., 2011). One option would be to build an optimal design thanks to the {`AlgDesign`} or the {`OptimalDesign`} packages that calculate the experimental designs for D, A, and I criteria. An example is given below in the case of a mixture design but would apply to regular factorial designs as well.

### 8.2.2 Mixture Designs

In many projects (e.g. in the food industry, in the personal care industry), optimizing a product's formula implies adjusting the proportions of its ingredients. In such cases, the proportions are interdependent (the total sum of all components of a mixture must be 100%). Therefore, these factors (the proportions) must be treated as mixture components. Mixture designs are usually represented using ternary diagrams.

The {mixexp} package offers a very convenient way to do this. In addition to creating the design, `DesignPoints()` allows to display the corresponding ternary diagram. Below is the example of a simplex-lattice design for three components and three levels obtained thanks to the function `SLD` (Figure 8.2):

```
library(mixexp)
mdes <- SLD(fac=3, lev=3)
DesignPoints(mdes)
```

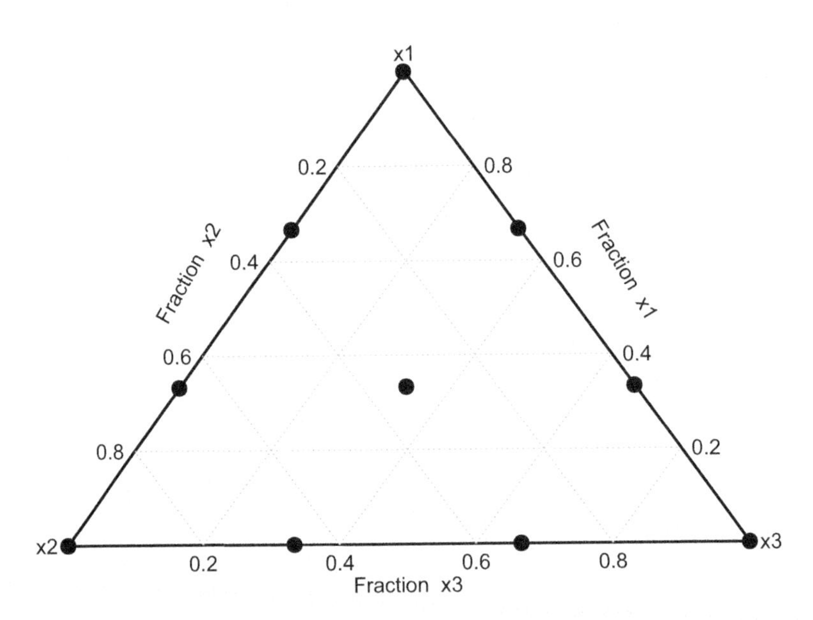


**FIGURE 8.2**
Ternary diagram of a simplex-lattice mixture design with three components.

Suppose that we want to adjust a biscuit recipe to optimize its sensory properties, we can design an experiment in which the proportion of ingredients vary. Let's play with butter, sugar, and flour. All three combined would account for 75% of the dough recipe and the remaining 25% would consist of other ingredients that we won't modify here (eggs, milk, chocolate, etc.). Besides, not any amount of these three ingredients would make sense (a biscuit with 75% of butter is not a biscuit, even in Brittany). We thus need to add constraints (ex: butter varies between 15 and 30% of this blend, sugar varies between 25 and 40%, and flour varies between 30 and 50%). Given this set of constraints (defined by `uc` for the upper contraints and `lc` for the lower constraints), we can use `mixexp::Xvert` to find the extreme vertices of our design (by also including a edge centroid using `ndm=1`):

```
mdes2 <- Xvert(nfac=3, uc=c(.30, .40, .50), lc=c(.15, .25, .30),
               ndm = 1, plot = FALSE) %>%
       mutate(across(where(is.numeric), round, digits = 3))
```

However, this design implies creating 11 mixtures, which is more than needed to apply a Scheffé quadratic model (Lawson and Willden, 2016). To reduce the number of mixtures and still allow fitting a quadratic model, we can use the `optFederov()` function from {`AlgDesign`} to select a D-optimal subset. Here, let's limit to nine products (`nTrials=9`) (Figure 8.3).

```
MixBiscuits <- optFederov(~ -1+x1+x2+x3+x1:x2+x1:x3+x2:x3+x1:x2:x3,
                          mdes2, nTrials=9)
DesignPoints(MixBiscuits$design, axislabs = c("Butter","Sugar","Flour"),
             pseudo = TRUE)
```

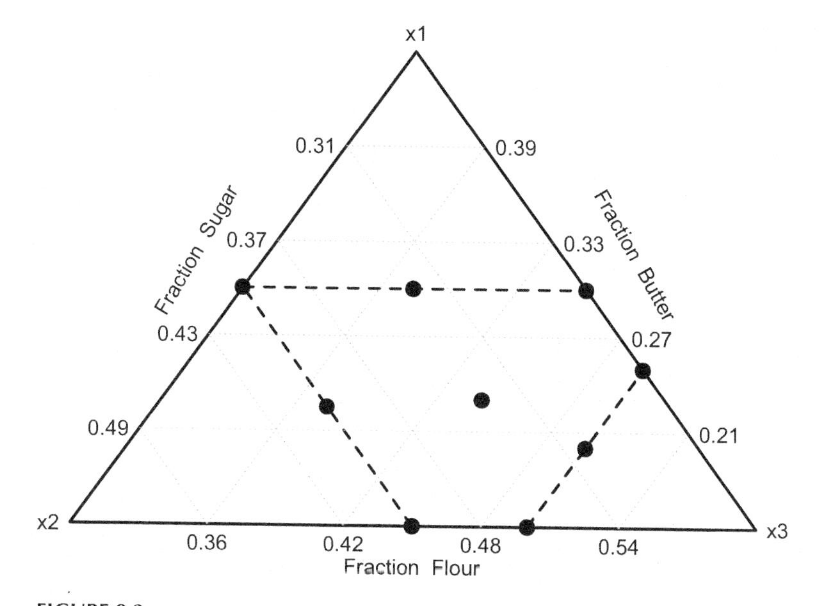


**FIGURE 8.3**
D-optimal mixture design with constrained factors.

Once the design is built, it could be desirable to randomize the order in which each sample is being made to avoid further biases. Suppose that we obtain average liking scores for our nine biscuits as given in Table 8.2 and stored in `Bmixt`:

```
Bmixt <- MixBiscuits$design %>%
  as_tibble() %>%
  dplyr::select(-4) %>%
  mutate(Product = LETTERS[1:9], .before=x1) %>%
  mutate(scores = c(7.5, 5.4, 5.5, 7.0, 6.0, 8.0, 5.8, 6.8, 7.9)) %>%
  rename("Butter"=x1, "Sugar"=x2, "Flour"=x3, "Liking"=scores)
```

**TABLE 8.2**

Average liking scores obtained for the biscuits from the mixture design

| Product | Butter | Sugar | Flour | Liking |
|---|---|---|---|---|
| A | 0.300 | 0.250 | 0.450 | 7.5 |
| B | 0.150 | 0.400 | 0.450 | 5.4 |
| C | 0.300 | 0.400 | 0.300 | 5.5 |
| D | 0.250 | 0.250 | 0.500 | 7.0 |
| E | 0.150 | 0.350 | 0.500 | 6.0 |
| F | 0.300 | 0.325 | 0.375 | 8.0 |
| G | 0.225 | 0.400 | 0.375 | 5.8 |
| H | 0.200 | 0.300 | 0.500 | 6.8 |
| I | 0.230 | 0.330 | 0.440 | 7.9 |

Once the data are collected we can use the `mixexp::MixModel()` function to fit a linear model and `mixexp::ModelPlot()` to draw a contour plot. This simple code would allow to get a contour plot that shows where would be the optimal area for the biscuit formulation (Figure 8.4).

---

Regardless of the construction of the mixture design, ternary diagrams are easy to plot with packages such as {`ggtern`} or {`Ternary`}. {`ggtern`} is particularly interesting since it builds on {`ggplot2`} and uses the same syntax.

---

```
invisible(
  capture.output(res <- MixModel(Bmixt, response="Liking",
                                 mixcomps=c("Butter","Sugar","Flour"),
                                 model=4)))
```

```
ModelPlot(model = res,
          dimensions = list(x1="Butter", x2="Sugar", x3="Flour"),
          lims = c(0.15,0.30,0.25,0.40,0.30,0.50), constraints = TRUE,
          contour = TRUE, cuts = 12, fill = TRUE, pseudo = TRUE,
          axislabs = c("Butter", "Sugar", "Flour"))
```

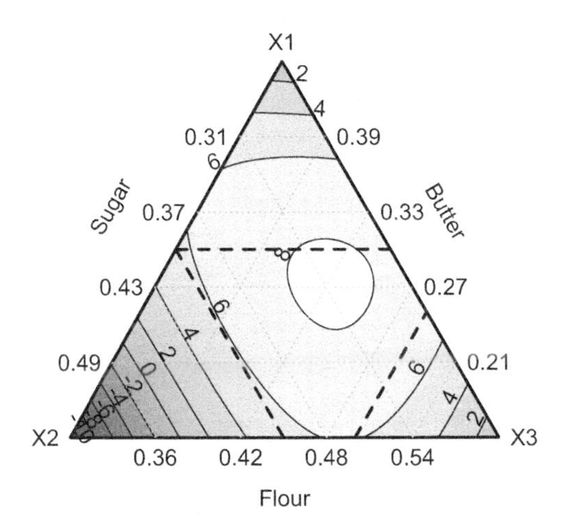


**FIGURE 8.4**
Modeled liking displayed as a contour plot on the ternary diagram.

According to these data, the optimal biscuit would have 31% of sugar, 27% of butter, and 42% of flour, and would reach a predicted liking score larger than 8.

### 8.2.3 Screening Designs

Product development is not a monolithic process and in the early stages of a project it could be extremely useful to use a design of experiment in combination with sensory evaluation to identify most influential factors of interest (Mao and Danzart, 2007; Pineau et al., 2019). Factorial and mixture designs belong to the product developers' essential toolkit and could serve this purpose. In practice however, they can only include a relatively limited number of factors. By contrast, fractional factorial designs (aka screening designs) are extremely efficient at dealing with many factors, pending some sacrifices on the estimation of interactions and quadratic effects. If, for example, we want to estimate the effect of 5 factors and assume that three- and four-factor interactions are negligible, we can then build a $2^{5-1}$ design (of, thus, 16 trials instead of 32) in which main effects are confounded with four-way interactions, and two-factor interactions are confounded with three-factor interactions. This design can be easily obtained with the {FrF2} package, with this simple command:

```
library(FrF2)
FrF2(nruns=16, nfactors=5, randomize=FALSE)
```

To reduce the number of trials even further, we can go as in the example below with a quarter fraction $2^{k-2}$ design, in which each effect that can be estimated is confounded with three other interactions.

```
FrF2(nruns=8, nfactors=5, randomize=FALSE)
```

Although fractional designs are only scarcely used, studies have shown that they could greatly contribute to sensory-led development of food (Baardseth et al., 2005; Modi and Prakash, 2008; Rytz et al., 2017; Pineau et al., 2019) as well as non-food product (Dairou et al., 2003).

For higher number of factors, Plackett-Burman designs are the most commonly used fractional factorial designs. They can be easily obtained with the `pb()` function of the {FrF2} package. For example, `FrF2::pb(12)` yields a 12-trial design that allows to test the effects of 11 factors.

### 8.2.4 Sensory Informed Designs for Product Development

Eventually, it is worth mentioning that, in some cases, sensory properties themselves can be used as factors and thus be implemented in a DoE. In this line of thinking, Naes and Nyvold have suggested that working this way would leave more scope for creativity (Naes and Nyvold, 2004). Naturally, this implies that product developers (1) have access to the measure of these properties and (2) can control the level of these properties and their interactions. These requirements are rarely met in food development but can be more easily implemented in some nonfood applications (see for example Petiot, 2022).

A specific application consists of using the sensory information available to make a selection of a subset of products, as described above (section 8.1.4).

## 8.3 Execute

Sir Ronald Fisher famously said in his presidential address to the first Indian statistical congress (1938): *"To consult the statistician after an experiment is finished is often merely to ask him to conduct a post-mortem examination. He can perhaps say what the experiment died of."*

Hopefully, the sections above would have helped sensory and consumer scientists designing their experiment in a way that would warrant them relevant and meaningful data that are obtained with maximum efficiency.

Fisher continues: *"To utilise this kind of experience the statistician must be induced to use his imagination, and to foresee in advance the difficulties and uncertainties with which, if they are not foreseen, his investigations will be beset."* Fortunately, we can spare the reader some of these imagination efforts and reiterate the fundamental principles of sensory evaluation that should help avoiding major pitfalls.[3]

- **Individual evaluation**
  Probably the most important requirement for the validity of sensory measurements is to perform individual evaluation. Sensory responses are very easily biased when judges can communicate. When this happens, observations cannot be considered independent which would rule out most statistical tests. Although this principle is generally accepted and correctly applied, some situations may be more challenging in this regard (such as project team meetings, b2b sample demonstration, tasting events, etc.). Individual evaluation is usually ensured by the use of partitioned sensory booths, but it can also be achieved by other means (table-top partitions, curtains, separate tables, or separate rooms). There are some cases, in consumer research, where interactions between subjects are allowed or even encouraged because they correspond to real-life situations. But these are exceptions to the rule, and in such cases, observations are to be considered at the group level.
- **Balanced order effects and treatments**
  We already discussed the importance of balancing the evaluation order for first-order and carry-over effects (Section 8.1.2). We cannot overstate how necessary this precaution is to get valid data. On top of having to deal with such effects, sensory scientists sometimes want to test how products are perceived (or liked) under different conditions (e.g. blind vs. branded, with/without nutritional information, in the lab vs. at home, etc.). Choice must then be made between a within-group design

[3] For a more detailed description of these principles, we refer the reader to comprehensive sensory evaluation textbooks. See for instance Lawless and Heymann (2010); Civille and Carr (2015); and Stone et al. (2020).

(in which participants evaluate the products under different conditions) and a between-group design (in which participants evaluate the product under one condition only). As often in consumer science, there is no perfect experiment and these two options have pros and cons. For instance, the within-group design would be more powerful and would allow data analysis at the individual level, but it would be more likely to induce response biases. Note that in both cases, participants must be randomly assigned to one group (corresponding either to a given condition or to the order in which each condition is being experienced if the study follows a within-group design).

- **Blind evaluation and controlled evaluation conditions**
  The primary goal of most sensory tests is to measure panelists' responses based on sensory properties only, without the interference of other variables that are seen as sources of potential biases. For this reason, tests are most frequently conducted on a blind-labeled basis without any information regarding the samples being tested (product identity, brand, price, nutritional facts, claims, etc.). Samples are thus usually blind labeled with random three-digit codes. This way, focus is placed on sensory perception and not on memory or expectations. Even information about the presence of duplicates or about the total number of products included in the design could induce biases. However, it is not always possible to hide all information (e.g., when the brand is printed directly on the product). It should also be noted that information is sometimes included as part of the study design to precisely evaluate the effect of that information. Besides, when sensory evaluation is used for market research goals, evaluation of the full mix can be preferred.

  Along the same lines and always in the view of collecting accurate and repeatable data, sensory scientists strive to control evaluation conditions. Sensory booths serve this purpose as they allow individual evaluation under controlled and standardized conditions. Nevertheless, for consumer tests (especially for hedonic tests), researchers may value the role of context in judgment construction and decision-making, and thus seek to contextualize their experimental setup for gains in ecological validity (Galiñanes Plaza et al., 2019).[4]

- **Separate affective from analytical tasks**
  For sensory evaluation, a clear distinction is usually made between analytical measurements (whereby emphasis is placed on description of sample characteristics or on differences and similarities between samples) and affective measurements (whereby focus is placed on liking, preferences, and emotions that may derive from the consumption of a

[4] Sensory and consumer research facilities such as living labs or immersive spaces are used in efforts to better account for the role of context without compromising on control. For more information on this topic, see Meiselman (2019).

product). Because the tasks involved in these two types of measurements are very different, the general recommendation is to conduct them separately (and most often, with different people). Proceeding otherwise would risk inducing cognitive biases and collecting skewed – or even meaningless – data. For example, if the goal of a study is to measure how much consumers like a given set of food products, it wouldn't make sense to ask trained panelists to rate their liking for the products they have been trained to describe. They can certainly do it, but their judgment of the products is likely to be changed by that training and by their extensive exposure to the product. Therefore, they can no longer be considered *normal* consumers. This is relatively commonsense. However, the risk of biases can sometimes be more subtle. Indeed, it might be tempting to ask consumers to give their liking for samples and, within the same session, to describe the same samples for a number of attributes. By doing so, you risk changing participants' mindset (e.g. by over-focusing on specific attributes) and thus altering liking scores (Popper et al., 2004 and Prescott et al., 2011). There is much debate though about which type of descriptive tasks would actually lead to biased responses (Jaeger et al., 2015). With this in mind, experimenters might still consider conducting combined measurements for product optimization, especially to get rough estimates of product specifications to target, in the first stage of product development. In this objective, Just-about-right (JAR) scales or the Ideal Profile Method (IPM) are very popular tasks. They provide a very direct way to optimize products' sensory characteristics (Rothman and Parker, 2009 and Worch et al., 2013). Alternately, one might expect that "untrained" consumers cannot be used for descriptive analysis. However in the past few decades, the development of descriptive methods that do not require training and that can be achieved in a single session has made consumer-based descriptive analysis possible, reliable, and endorsed (Varela and Ares, 2012; Ares and Varela, 2017; and Pineau et al., 2022).

- **Sample availability**
  An obvious, but essential, condition for conducting sensory evaluation is to have samples available for testing. It is surprising to see how many sensory studies fail simply because the experimenters have not anticipated the production of experimental samples in sufficient quantities or procurement of commercial products. Especially, remember that for many sensory tests, samples are needed for training in addition to the evaluation itself. No data analysis can make up for a lack of samples, no matter how sophisticated it may be. We therefore strongly advise experimenters to review their need for samples when they design a study and, if they do not make the samples themselves, to discuss with their clients or project teams (R&D, pilot plant, suppliers, etc.) to ensure that samples will be available over the course of the study.

- **Regulations for studies with human subjects**
  Running a sensory or a consumer study implies working with human subjects at some point (online surveys and simple passive observation count!). Therefore, experimenters must ensure that their protocol complies with local and international rules. Most often, research projects should be approved by an Institutional Review Board (IRB) or an appropriate ethical committee. As far as data are concerned, it is also important to ensure that data collection, use, and storage comply with applicable regulations such as EU's General Data Protection Regulation (GDPR) or the California Consumer Privacy Act (CCPA).
- **Quantification**
  Finally, it is critical that sensory and consumer scientists anticipate what type of analysis they will conduct in accordance with the exact information they are looking for and thus define what data type and scaling method they will adopt (see OMahony, 1986, and Lawless and Heymann, 2010). The means of quantification (counts, sorting, ranking, scaling, mapping, reaction time, etc.) has usually been set long before execution, when the study was designed. When time comes to run the tests, the experimenter will have to rely on a proper and reliable way to collect data. Nowadays, commercial sensory software solutions allow to collect any type of data, including temporal information. However, in some cases, the experimenter may choose to ask panelists to use paper and pencil, or just to give a verbal answer, or to arrange the samples physically on a bench. Care must then be taken to ensure proper coding scheme and data entry. At this stage, it is important to keep as much information as possible on the experimental details, such as who evaluated which sample, in which order, at what time, etc. It is usually advised to try to enter data in a single spreadsheet with one column per variable and one line per observation, but in some rare cases, it might be more convenient to enter each panelist's data in separate tabs. This could be the case, for example, for methods like free sorting or napping. Note that entering data is prone to mistakes and typos, especially when entered manually into a spreadsheet. In the next sections, we will see how to import data from that spreadsheet into R (Section 8.4) and how to check for outliers and missing values (Section 9.2.2).

## 8.4 Import

It is a truism, but to analyze data we first need *data*. If the data are already available in R, then the analysis can be performed directly. However, in most cases, the data are stored outside the R environment and need to be imported.

In practice, the data might be stored in as many formats as one can imagine, despite how it ends up being a fairly common solution (*.txt* file, *.csv* file, or *.xls(x)* file) or software specific (e.g. Stata, SPSS, etc.).

Since it is very common to store the data in Excel spreadsheets (*.xls(x)*) due to its simplicity, the emphasis is on this solution. Fortunately, most generalities presented for Excel files also apply to other formats through `base::read.table()` for *.txt* files, `base::read.csv()` and `base::read.csv2()` for *.csv* files or through the {`read`} package (which is part of the {`tidyverse`}).

For other (less common) formats, you may find alternative packages that would allow importing your files in R. Particular interest can be given to the package {`rio`} (*rio* stands for *R* *I*nput and *O*utput) which provides an easy solution that:

1. Handles a large variety of files,
2. Guess the type of file it is,
3. Provides tools to import, export, and convert almost any type of data format, including *.csv*, *.xls(x)*, or data from other statistical software such as SAS (*.sas7bdat* and *.xpt*), SPSS (*.sav* and *.por*), or Stata (*.dta*).

Similarly, the package {`foreign`} provides functions that allow importing data stored from other statistical software (incl. Minitab, S, SAS, Stata, SPSS, etc.).

Although Excel is most likely one of the most popular way of storing data, there are no {`base`} functions that allow importing such files directly. Fortunately, many packages have been developed for that purpose, including {`XLConnect`}, {`xlsx`}, {`gdata`}, and {`readxl`}. Due to its convenience and speed of execution, we will focus on {`readxl`}.

### 8.4.1 Importing Structured Excel File

First, let's import the *biscuits_sensory_profile.xlsx* workbook using `readxl::read_xlsx()` by informing as parameter the `location` of the file and the `sheet` where it is stored. For convenience, we are using the {`here`}[5] package to retrieve the path of the file (stored in `file_path`).

This file is said to be *structured* as all the relevant information is already stored in the same sheet in a structured way. In other words, no decoding is required here, and there are no "unexpected" rows or columns (e.g. empty lines, or lines with additional information regarding the data that is not data):

[5] The package {`here`} is very handy as it provides an easy way to retrieve your file's path (within your working directory) by simply giving the name of the file and folder in which they are stored in.

- The first row within the *Data* sheet of *biscuits_sensory_profile.xlsx* contains the headers;
- From the second row onward, only data are being stored.

Since this data will be used for some analyses, it is assigned data to an R object called `sensory`.

```
library(here)
file_path <- here("data","biscuits_sensory_profile.xlsx")

library(readxl)
sensory <- readxl::read_xlsx(file_path, sheet="Data")
```

To ensure that the importation went well, it is advised to print `sensory` after importation. Since {`readxl`} has been developed by Hadley Wickham and colleagues, its functions follow the {`tidyverse`} principles and the data thus imported are stored in a `tibble`. Let's take advantage of the printing properties of a `tibble` to evaluate `sensory`:

```
sensory
```

```
## # A tibble: 99 x 34
##   Judge Product Shiny Externa~1 Color~2 Qty o~3 Surfa~4
##   <chr> <chr>   <dbl>     <dbl>   <dbl>   <dbl>   <dbl>
## 1 J01   P01      52.8      30      22.8     9.6    22.8
## 2 J01   P02      48.6      30      13.2    10.8    13.2
## 3 J01   P03      48        45.6    17.4     7.8    14.4
## 4 J01   P04      46.2      45.6    37.8     0      48.6
## # ... with 95 more rows, 27 more variables:
## #   'Print quality' <dbl>, Thickness <dbl>,
## #   'Color contrast' <dbl>,
## #   'Overall odor intensity' <dbl>,
## #   'Fatty odor' <dbl>, 'Roasted odor' <dbl>,
## #   'Cereal flavor' <dbl>, 'RawDough flavor' <dbl>,
## #   'Fatty flavor' <dbl>, 'Dairy flavor' <dbl>, ...
```

`sensory` is a tibble with 99 rows and 34 columns that includes the `Judge` information (first column, defined as character), the `Product` information (second column, defined as character), and the sensory attributes (third column onward, defined as numerical or `dbl`).

### 8.4.2 Importing Unstructured Excel File

In some cases, the data are not so well organized/structured, and may need to be *decoded*. This is the case for the workbook entitled *biscuits_traits.xlsx*.

In this file:

- The variables' name have been coded and their corresponding names (together with some other valuable information we will be using in Chapter 10) are stored in a different sheet entitled *Variables*;
- The different levels of each variable (including their code and corresponding names) are stored in another sheet entitled *Levels*.

To import and decode this data set, multiple steps are required:

1. Import the variables' name only;
2. Import the information regarding the levels;
3. Import the data without the first line of header, but by providing the correct names (obtained in the step 1);
4. Decode each question (when needed) by replacing the numerical code with their corresponding labels.

Let's start with importing the variables' names from *biscuits_traits.xlsx* (sheet *Variables*)

```
file_path <- here("data","biscuits_traits.xlsx")
var_names <- readxl::read_xlsx(file_path, sheet="Variables")
```

```
## # A tibble: 62 x 5
##   Code  Name        Direction Value `Full Question`
##   <chr> <chr>       <chr>     <dbl> <chr>
## 1 Q1    Living area <NA>         NA <NA>
## 2 Q2    Housing     <NA>         NA <NA>
## 3 Q3    Judge       <NA>         NA <NA>
## 4 Q4    Height      <NA>         NA <NA>
## # ... with 58 more rows
```

In a similar way, let's import the information related to the levels of each variable, stored in the *Levels* sheet. A deeper look at the *Levels* sheet shows that only the coded names of the variables are available. In order to include the final names, `var_names` is joined (using `inner_join`).

```
var_labels <- readxl::read_xlsx(file_path, sheet="Levels") %>%
  inner_join(dplyr::select(var_names, Code, Name),
             by=c(Question="Code"))
```

```
## # A tibble: 172 x 4
##   Question  Code Levels       Name
##   <chr>    <dbl> <chr>        <chr>
## 1 Q1           1 Urban Area   Living area
## 2 Q1           2 Rurban Area  Living area
## 3 Q1           3 Rural Area   Living area
## 4 Q2           1 Apartment    Housing
## # ... with 168 more rows
```

Ultimately, the data (*Data*) are imported by substituting the coded names with their corresponding names. This process can be done by skipping reading the first row of the data that contains the coded header (`skip=1`), and by passing `Var_names` as header or column names (after ensuring that the names' sequence perfectly match across the two tables!).

Alternatively, you can import the data by specifying the range in which the data is being stored (here 'range="A2:BJ108" ").

The data now have the right headers, however, each variable is still coded numerically. This step to convert the numerical values with their corresponding labels is described in Chapter 9.

---

Datasets sometimes include a sub-header with extra information regarding the levels of a factor. In such a case, a similar approach should be used: 1. Start with importing the first $n$ rows of the data that contain this information using the parameter `n_max` from `readxl::read_xlsx`. 2. From this subset, extract the column names. 3. For each variable (when information is available), store the additional information as a list of tables that contain the code and their corresponding label. 4. Reimport the data by skipping these $n$ rows and by manually informing the headers.

---

### 8.4.3 Importing Data Stored in Multiple Sheets

Sometimes, the data to be analyzed are stored in different files or in different sheets within the same file. This would typically occur if a test involving the same samples is repeated over time, or has been run simultaneously in different locations, or simply for convenience for the person who manually collected the data.

Since the goal here is to highlight the possibilities in R to handle such situations, we propose to use a small fake example where 12 panelists evaluated 2 samples on 3 attributes in 3 sessions, each session being stored in a different sheet in *excel_scrap.xlsx*.

A first approach to tackle this problem could be to import each file separately and to combine them together using the `bind_rows()` function from the {`dplyr`} package. However, this solution is not optimal since it is very tedious when a larger number of sheets is involved, and it is not automated since the code will no longer run (or be incomplete) when the number of sessions changes.

Instead, we prefer to fully automate the importation. To do so, let's first introduce `excel_sheets()` from {`readxl`}: this function provides the name of all the sheets that are available in the file of interest in a list. Then, through `map()` from the {`purrr`} package, we apply `read_xlsx()` to all the elements one by one of obtained with `excel_sheets()`.

```
path <- file.path("data", "excel_scrap.xlsx")
files <- path %>%
  excel_sheets() %>%
  set_names(.) %>%
  map(~readxl::read_xlsx(path, sheet = .))
```

As can be seen, this procedure creates a list of tables, with as many elements as there as sheets in the excel file.

---

As an alternative, consider using `import_list()` from {`rio`} as it imports automatically all the sheets from a spreadsheet with one single command.

---

To convert this list of data tables into one unique data frame, we first extend the previous code and `enframe()` it by informing that the separation was based on `Session`. Once done, the data (stored in `data`) are still nested in a list and should be *unfolded*. Such operation is done with the `unnest()` function from {`tidyr`}:

```
files %>%
  enframe(name = "Session", value = "data") %>%
  unnest(cols = c(data))
```

```
## # A tibble: 72 x 6
##   Session   Subject Sample Sweet  Sour Bitter
##   <chr>     <chr>   <chr>  <dbl> <dbl>  <dbl>
## 1 Session 1 J1      P1     46.6   82.6   25.5
## 2 Session 1 J2      P1      1.28  60.1   13.9
## 3 Session 1 J3      P1     29.1   48.5   62.8
## 4 Session 1 J4      P1     29.9   79.2   52.7
## # ... with 68 more rows
```

This procedure finally returns a tibble with 72 rows and 6 columns, ready to be analyzed!

---

Few additional remarks regarding the last set of code: 1. Instead of `enframe()`, we could have used `reduce()` from {`purrr`}, or `map()` combined with `bind_rows()`. However, both these solutions have the drawbacks that the information regarding the `Session` would be lost since it is not part of the data set itself. 2. The functions `enframe()` and `unnest()` have their alter-ego in `deframe()` and `nest()` which aim at transforming a data frame into a list of tables and in nesting data by creating a list-column of data frames. 3. In case the different sets of data are stored in different excel files (rather than different sheets within a file), we could apply a similar procedure by using `list.files()` (instead of `excel_sheets()`) from the {`base`} package, together with `pattern = "xlsx"` to limit the search to Excel files present in a predefined folder. Such a solution becomes handy when many similarly structured files are stored in the same folder and need to be combined.

---

# 9

# *Data Preparation*

After importing the data, the next crucial step is to ensure that the data as they are now available are of good quality and are the correct representation of reality. As an example, during importation, software (such as R) tends to guess (from reading the file) the nature of each variable. If such guess is correct in 99% of cases, there are situations in which it is erroneous. Ignoring such errors can have huge consequences on the final results and conclusions. The goal of this section is hence to perform some pre-check of the data and to prepare them for future analyses.

## 9.1 Introduction

*Data Preparation*, which consists of *data inspection* and *data cleaning*, is a critical step before any further *Data Manipulation* or *Data Analysis*. Having a good data preparation procedure ensures a good understanding of the data and avoids what could be very critical mistakes.

To illustrate the importance of the later point, let's imagine a study in which the samples are defined by their three-digit code. During importation, R would recognize them as number and hence define the *Product* column as numerical. Without inspection and correction, any ANOVA that include the product effect would be replaced by a linear regression (or analysis of covariance) which of course does not provide the results required (although the analysis would run without error). Worst, if this procedure is automated and the *p-value* associated to the product effect is extracted, the conclusions would rely on the wrong analysis! A good data preparation procedure is therefore essential to avoid such unexpected results.

So what does data preparation consist of and how does it differ from data manipulation? There is clearly a thin line between data preparation (and particularly *data cleaning*) and data manipulation, as both these steps share many procedures (same applies to data manipulation and data analysis for

DOI: 10.1201/9781003028611-9

instance). Although there are multiple definitions for each step, we decided to follow the following rule:

*Data Preparation* includes all the steps required to ensure that the data match their intrinsic nature. These steps include inspecting the data at hand (usually through simple descriptive statistics of the data as a whole) and cleaning the data by eventually correcting importation errors (including the imputation of missing data). Although some descriptive statistics are being produced for data inspection, these analyses have no interpretation value besides ensuring that the data are in the right range or following the right distribution. For instance, with our sensory data, we would ensure that all our sensory scores are included between 0 and 100 (negative scores would not be permitted) but we would not look at the mean or the distribution of the score per product which would belong to data analyses as it would often lead to interpretation (e.g. P01 is sweeter than P02).

*Data Manipulation* is an optional step that adjusts or converts the data into a structure that is usable for further analysis. This, of course, may lead to *interpretation* of the results as it may involve some analyses.

The *Data Analysis* step ultimately converts the data into results (through values, graphics, tables, etc.) that provide more insights (through interpretation) about the data.

The data used in this chapter correspond to the *biscuits_sensory_profile.xlsx* that you already imported in Chapter 8 but have few missing values. This new data set is stored in *biscuits_sensory_profile_with_NA.xlsx*.

As usual, we start this chapter by loading the main packages we need and by importing this data set:

```
library(tidyverse)
library(readxl)
library(here)

file_path <- here("data","biscuits_sensory_profile_with_NA.xlsx")
sensory <- read_xlsx(file_path, sheet="Data")
```

## 9.2 Inspect

### 9.2.1 Data Inspection

To inspect the data, different steps can be used. First, since `read_xlsx()` returns a tibble, let's take advantage of its printing properties to get a feel of the data:

```
sensory
```

```
## # A tibble: 99 x 34
##   Judge Product Shiny Externa~1 Color~2 Qty o~3 Surfa~4
##   <chr> <chr>   <dbl>    <dbl>   <dbl>   <dbl>   <dbl>
## 1 J01   P01      52.8     30      22.8     9.6    22.8
## 2 J01   P02      48.6     30      13.2    10.8    13.2
## 3 J01   P03      48       45.6    17.4     7.8    14.4
## 4 J01   P04      46.2     45.6    37.8     0      48.6
## # ... with 95 more rows, 27 more variables:
## #   'Print quality' <dbl>, Thickness <dbl>,
## #   'Color contrast' <dbl>,
## #   'Overall odor intensity' <dbl>,
## #   'Fatty odor' <dbl>, 'Roasted odor' <dbl>,
## #   'Cereal flavor' <dbl>, 'RawDough flavor' <dbl>,
## #   'Fatty flavor' <dbl>, 'Dairy flavor' <dbl>, ...
```

Other informative solutions consist of printing a summary of the data through `summary()` or `glimpse()`:

```
summary(sensory)
```

```
##     Judge              Product              Shiny
##  Length:99          Length:99          Min.   : 0.0
##  Class :character   Class :character   1st Qu.: 9.3
##  Mode  :character   Mode  :character   Median :21.0
##                                        Mean   :23.9
##                                        3rd Qu.:38.4
##                                        Max.   :54.0
##  External color intensity Color evenness
##  Min.   : 6.6             Min.   : 6.6
##  1st Qu.:27.0             1st Qu.:19.5
##  Median :34.8             Median :26.4
##  Mean   :33.7             Mean   :28.2
##  3rd Qu.:42.6             3rd Qu.:37.2
##  Max.   :55.2             Max.   :53.4
##  Qty of inclusions
##  Min.   : 0.0
##  1st Qu.:13.8
##  Median :19.8
##  Mean   :20.6
##  3rd Qu.:29.1
##  Max.   :40.8
```

```
glimpse(sensory)
```

```
## Rows: 99
## Columns: 10
## $ Judge                        <chr> "J01", "J01", "J01~
## $ Product                      <chr> "P01", "P02", "P03~
## $ Shiny                        <dbl> 52.8, 48.6, 48.0, ~
## $ `External color intensity` <dbl> 30.0, 30.0, 45.6, ~
## $ `Color evenness`             <dbl> 22.8, 13.2, 17.4, ~
## $ `Qty of inclusions`          <dbl> 9.6, 10.8, 7.8, 0.~
## $ `Surface defects`            <dbl> 22.8, 13.2, 14.4, ~
## $ `Print quality`              <dbl> 48.6, 54.0, 49.2, ~
## $ Thickness                    <dbl> 38.4, 35.4, 25.8, ~
## $ `Color contrast`             <dbl> 37.8, 40.2, 17.4, ~
```

These functions provide basic yet relevant views of each variable present in the data, including their types, the range of values, means, and medians, as well as the first values of each variable.

Such a view might be sufficient for some first conclusions (e.g. Are my panelists considered as numerical or nominal data? Do I have missing values?), yet it is not sufficient to fully ensure that the data are ready for analysis. For the latter, more extensive analyses can be performed automatically in different ways. These analyses include looking at the distribution of some variables or the frequencies of character levels.

A first solution comes from the {`skimr`} package and its `skim()` function. When applying it to data, an automated extended summary is directly printed on screen by separating `character` type variables from `numeric` type variables:

```
library(skimr)
skim(sensory)
```

Another approach consists of generating automatically an html report with some predefined analyses using `create_report()` from the {`DataExplorer`} package.

```
library(DataExplorer)
create_report(sensory)
```

Unless specified otherwise through `output_file`, `output_dir`, and `output_format`, the report will be saved as an html file on your active directory as *report.html*. This report provides many statistics on your data, including some simple statistics (e.g. raw counts, percentages), informs you on the structure of your data, as well as on potential missing values. It also generates graphics to describe your variables (e.g. univariate distribution, correlations, and PCA).

---

Note that the analyses performed to build this report can be called directly within R. For instance, `introduce()` and `plot_intro()` generate the first part of the report, whereas `plot_missing()` and `profile_missing()` provide information regarding missing data just to name those.

---

### 9.2.2 Missing Data

From the data inspection described in Section 9.2.1, it can be seen that the data set contains missing values. For instance, the attribute `Light` has one detected missing value. There are different ways in which we can handle such missing values. But first, let's try to find out where these missing values are and what impact they may have on our analyses (are they structured or unstructured, etc.)

#### *Visualization of Missing Values*

A first approach to inspect and visualize where the missing values are is by representing them visually. To do so, the {`visdat`} package provides a neat solution as it represents graphically the data by highlighting where missing values are located (Figure 9.1). Such visual representation is obtained using the `vis_miss()` function:

```
library(visdat)
sensory %>%
  vis_miss()
```

As can be seen, missing values are only present in few variables. However, `Sour` contains up to 10% of missing data, which can be quite critical in some situations.

If we would want to dig deeper and assess for which (say) products data are missing, we could recreate the same plots per product. The following code would generate that for you:

```
sensory %>%
  split(.$Product) %>%
  map(function(data){
    vis_miss(data)
  })
```

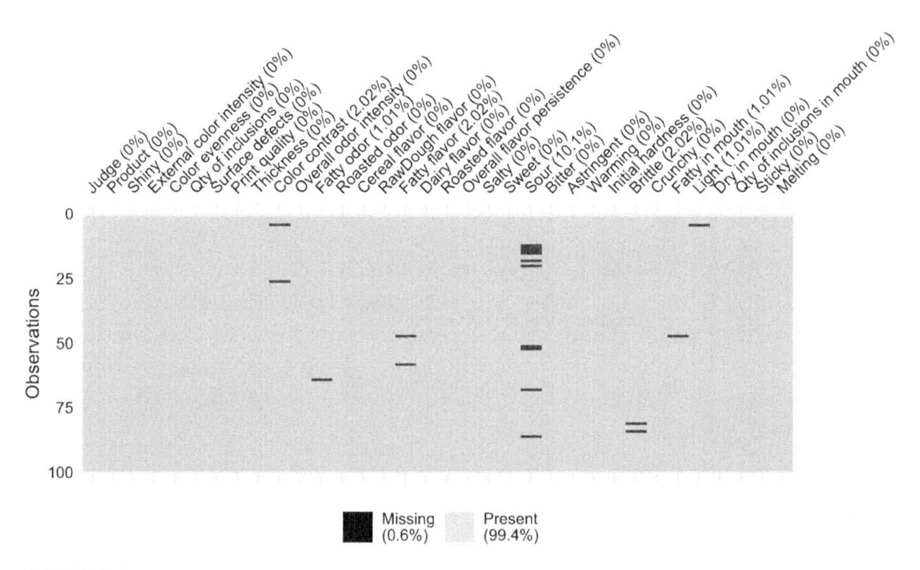


**FIGURE 9.1**
Visualization of missing values in the sensory data set using visdat::vismiss().

Of course, this approach could also be applied per panelist for instance.

Once we have investigated where the missing values are located, we can go further by understanding if there are some sorts of relationship between missing values. In other words, are the missing values random? Or are they somewhat structured? To answer these questions, the {`naniar`} package provides an interesting function called `gg_miss_upset()` which studies the relationship between missing values (Figure 9.2):

```
library(naniar)
sensory %>%
  gg_miss_upset()
```

It seems here that the only connection between NAs is observed between `Light` and `Color contrast`.

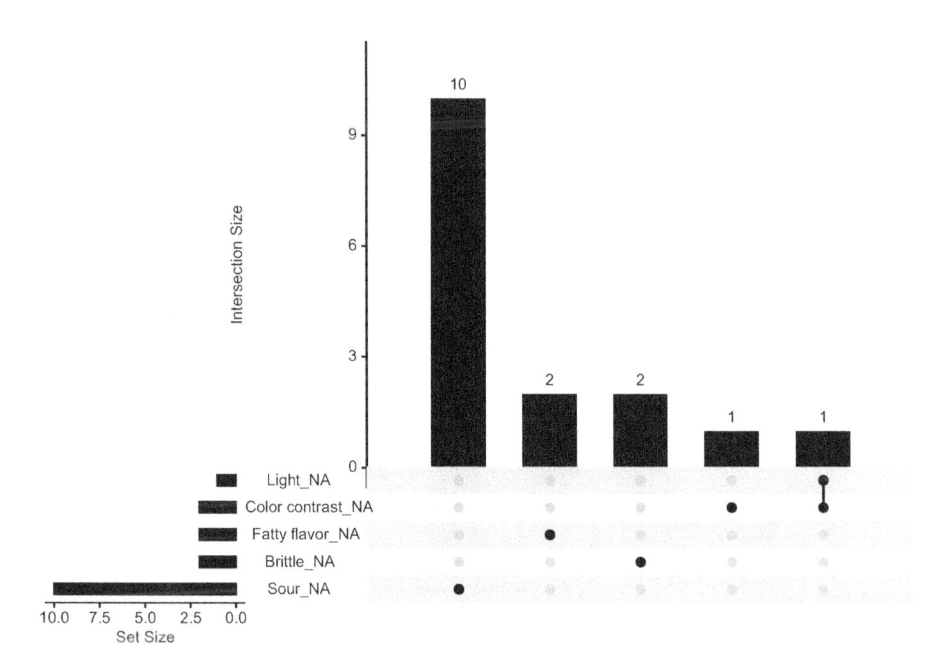


**FIGURE 9.2**
Visualization of the pattern of missing values in the sensory data thanks to naniar::gg_miss_upset().

Such a relational structure can also be visualized in a scatter plot using the `geom_miss_point()` function from the same package (Figure 9.3):

```
ggplot(sensory, aes(x=Product, y=Sour))+
  geom_miss_point()
```

Here, the relationship between `Product` and `Sour` is shown. Such a plot may help decide what to do with missing values, whether it is ignoring, removing, or predicting them.

### *Ignoring Missing Values*

A first solution to handle missing values is to simply *ignore* them, as many analyses handle them well. For instance, an ANOVA could be run for such attributes, and results are being produced:

```
broom::tidy(aov(Light ~ Product + Judge, data=sensory))
```

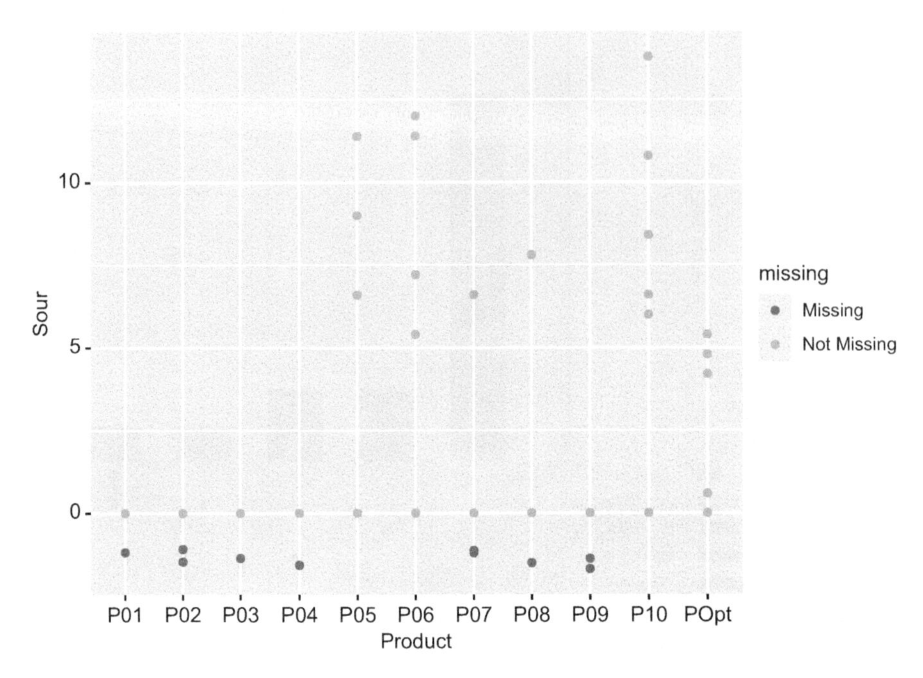


**FIGURE 9.3**
Visualization of missing values for Sour with geom_miss_point().

```
## # A tibble: 3 x 6
##   term          df sumsq meansq statistic  p.value
##   <chr>      <dbl> <dbl>  <dbl>     <dbl>    <dbl>
## 1 Product       10 2379.  238.       4.73  2.71e-5
## 2 Judge          8 4100.  513.      10.2   1.16e-9
## 3 Residuals     79 3975.   50.3     NA    NA
```

This solution may work fine when the number of missing values is small, but be aware that it can also provide erroneous results in case they are not handled the way the analyst expects them to be handled.

For some other analyses, *ignoring* the presence of missing values may simply provide unwanted results. To illustrate this, let's compute the simple mean per product for `Light`

```
sensory %>%
  group_by(Product) %>%
  summarise(Light = mean(Light)) %>%
  ungroup()
```

```
## # A tibble: 11 x 2
##   Product Light
##   <chr>   <dbl>
## 1 P01      29.6
## 2 P02      30.9
## 3 P03      28.3
## 4 P04      NA
## # ... with 7 more rows
```

As can be seen, missing values for `P04` are conducive to a mean defined as `NA`.

### *Removing Missing Values*

To force the mean to be computed, we must inform R to remove any observation containing missing values beforehand. Such procedure can be done manually by simply filtering out any missing data (here for `Sour`) before running the analysis:

```
sensory %>%
  filter(!is.na(Sour))
```

```
## # A tibble: 89 x 34
##   Judge Product Shiny Externa~1 Color~2 Qty o~3 Surfa~4
##   <chr> <chr>   <dbl>     <dbl>   <dbl>   <dbl>   <dbl>
## 1 J01   P01      52.8      30      22.8     9.6    22.8
## 2 J01   P02      48.6      30      13.2    10.8    13.2
## 3 J01   P03      48        45.6    17.4     7.8    14.4
## 4 J01   P04      46.2      45.6    37.8     0      48.6
## # ... with 85 more rows, 27 more variables:
## #   'Print quality' <dbl>, Thickness <dbl>,
## #   'Color contrast' <dbl>,
## #   'Overall odor intensity' <dbl>,
## #   'Fatty odor' <dbl>, 'Roasted odor' <dbl>,
## #   'Cereal flavor' <dbl>, 'RawDough flavor' <dbl>,
## #   'Fatty flavor' <dbl>, 'Dairy flavor' <dbl>, ...
```

However, this latter solution is not always satisfactory as it also deletes actual data. Indeed, as a result, the data set went from 99 to 89 rows. This means that existing data were removed for corresponding observations, even for variables that did not have missing values.

Therefore, we prefer another alternative which consists in removing missing values within the analysis procedure (here `mean()`) through the parameter `na.rm=TRUE`:

```
sensory %>%
  group_by(Product) %>%
  summarise(Light = mean(Light, na.rm=TRUE)) %>%
  ungroup()
```

```
## # A tibble: 11 x 2
##   Product Light
##   <chr>   <dbl>
## 1 P01      29.6
## 2 P02      30.9
## 3 P03      28.3
## 4 P04      40.7
## # ... with 7 more rows
```

Using `na.rm=TRUE` is equivalent to removing the missing values from the data before performing the analysis, but only for the variable of interest. A similar approach consists in first rotating (using `pivot_longer()`) the data before removing missing values:

```
sensory %>%
  pivot_longer(Shiny:Melting,
               names_to="Variables", values_to="Scores") %>%
  filter(!is.na(Scores)) %>%
  group_by(Product, Variables) %>%
  summarize(Means = mean(Scores)) %>%
  ungroup() %>%
  pivot_wider(names_from = Variables, values_from = Means) %>%
  dplyr::select(Product, Sour, Light)
```

```
## `summarise()` has grouped output by 'Product'. You can
## override using the `.groups` argument.
```

```
## # A tibble: 11 x 3
##   Product  Sour Light
##   <chr>   <dbl> <dbl>
## 1 P01         0  29.6
## 2 P02         0  30.9
## 3 P03         0  28.3
## 4 P04         0  40.7
## # ... with 7 more rows
```

This solution seems satisfactory as the means were computed without using `na.rm=TRUE` for both `Sour` and `Light` (who contained missing values). However, its use is limited since converting the data to its original format (i.e. performing `pivot_wider()` after `pivot_longer()` without computing the mean in between) will reintroduce the missing values.[1]

It should be noted that a consequence of removing missing values is data unbalance. For example, we can observe this for `Light` and `Sour`, by printing the number of panelists who evaluated each product:

```
sensory %>%
  pivot_longer(Shiny:Melting, names_to="Variables", values_to="Scores") %>%
  filter(!is.na(Scores),
         Variables %in% c("Light","Sour")) %>%
  group_by(Product, Variables) %>%
  count() %>%
  ungroup() %>%
  pivot_wider(names_from=Variables, values_from=n)
```

```
## # A tibble: 11 x 3
##   Product Light  Sour
##   <chr>   <int> <int>
## 1 P01         9     8
## 2 P02         9     7
## 3 P03         9     8
## 4 P04         8     8
## # ... with 7 more rows
```

Here, for example, the only missing value detected for `Light` is related to `P04`. For `Sour`, `P02`, `P07`, and `P09` only have seven observations out of nine.

The solution of *blindly* removing missing values is a solution that you may sometime use. However, it is not the only strategy, and we can consider other approaches that are more in-line with the nature of the data.

Rather than removing the missing values only, we could consider removing blocks of data, whether it is attributes, products, or panelists that present missing data. This solution is particularly handy when tests are performed in multiple sessions and some respondents did not manage to attend them all. It can then be relevant to remove completely those respondents from your data.

The procedure presented below show the procedure on how to remove attributes with missing data, but could easily be adapted to panelists or products:

[1] Missing values do not need to be visible to exist: Incomplete designs are a good example showing that although the data do not have empty cells, it does contain a lot of missing data (the samples that were not evaluated by each panelist).

```
sensory_long <- sensory %>%
  pivot_longer(Shiny:Melting,
               names_to="Variables", values_to="Scores")

attr_rmv <- sensory_long %>%
  filter(is.na(Scores)) %>%
  pull(Variables) %>%
  unique()

sensory_clean <- sensory_long %>%
  filter(!(Variables %in% attr_rmv)) %>%
  pivot_wider(names_from=Variables, values_from=Scores)
```

This procedure removed the 7 attributes that contained missing values (and stored in `attr_rmv`), leading to a table with 99 rows and 29 columns (instead of 36).

#### *Imputing Missing Values*

Rather than removing missing data, another strategy consists of imputing missing values. Here again, many strategies can be considered, starting with replacing them with a fixed value. Such an approach is usually not the most suitable one, yet it can be relevant in certain cases. For instance, in a CATA task, missing values are often replaced with 0s (not ticked).

To replace missing values with a fixed value, `replace_na()` can be used. When applied to a tibble, this function requires you defining which columns to apply it to and which values to use (each column being treated separately) using `list()`.

For convenience, let's apply it to `sensory` by replacing missing values for `Sour` by the value `888` and for `Light` with `999` (we use these extreme values to track changes more easily):

```
sensory %>%
  replace_na(list(Sour = 888, Light = 999)) %>%
  dplyr::select(Judge, Product, Sour, Light)
```

```
## # A tibble: 99 x 4
##   Judge Product  Sour Light
##   <chr> <chr>   <dbl> <dbl>
## 1 J01   P01         0  22.8
## 2 J01   P02         0  21
## 3 J01   P03         0  20.4
## 4 J01   P04         0 999
## # ... with 95 more rows
```

When dealing with intensity scales, it is more frequent to replace missing values by the mean score for that product and attribute. When the test is duplicated, the mean provided by the *panelist* × *product* × *attribute* combination across the different repetitions available is even preferred as it maintains individual variability within the scores.

This approach is a two-step process:

1. Compute the mean (since we do not have duplicates, we use the mean per product);
2. Combine it to the data.

For simplicity, `sensory_long` is used as the starting point:

```
prod_mean <- sensory_long %>%
  group_by(Product, Variables) %>%
  summarize(Mean = mean(Scores, na.rm=TRUE)) %>%
  ungroup()
```

```
## `summarise()` has grouped output by 'Product'. You can
## override using the `.groups` argument.
```

```
sensory_long %>%
  full_join(prod_mean, by=c("Product","Variables")) %>%
  mutate(Scores = ifelse(is.na(Scores), Mean, Scores)) %>%
  dplyr::select(-"Mean") %>%
  pivot_wider(names_from=Variables, values_from=Scores) %>%
  dplyr::select(Judge, Product, Sour, Light)
```

```
## # A tibble: 99 x 4
##   Judge Product  Sour Light
##   <chr> <chr>   <dbl> <dbl>
## 1 J01   P01         0  22.8
## 2 J01   P02         0  21
## 3 J01   P03         0  20.4
## 4 J01   P04         0  40.7
## # ... with 95 more rows
```

As can be seen, the missing value associated to `J01` for `Light` and `P04` has been replaced by `40.7`. In fact, any missing values related to `P04` and `Light` would automatically be replaced by `40.7` here. For other products (and other attributes), their respective means would be used.

When the model used to impute missing values is fairly simple (here, replacing by the mean corresponds to a simple 1-way ANOVA), the imputation can be done directly through the `impute_lm()` function from the {`simputation`} package. To mimic the previous approach, the one-way ANOVA is being used.[2] Here, missing data for both `Sour` and `Light` are being imputed independently using the same model:

```
library(simputation)

sensory %>%
  impute_lm(Sour + Light ~ Product) %>%
  dplyr::select(Judge, Product, Sour, Light)
```

```
## # A tibble: 99 x 4
##   Judge Product  Sour Light
##   <chr> <chr>   <dbl> <dbl>
## 1 J01   P01         0  22.8
## 2 J01   P02         0  21
## 3 J01   P03         0  20.4
## 4 J01   P04         0  40.7
## # ... with 95 more rows
```

As can be seen, this procedure provides the same results as before, but with fewer steps!

In some situations, implementing missing values using such ANOVA (or regression) model can lead to aberrations. It is for instance the case when the imputed values falls outside the scale boundaries. To avoid such situations, {`simputation`} also provides other more advanced alternatives including (among others) `impute_rf()` which uses random forest to impute the missing values.

Last but not least, imputation of missing values could also be done in a multivariate way, by using the structure of the data (e.g. correlation) to predict the missing values. This is the approach proposed in the {`missMDA`} package. Since our data are numeric, the imputation is done through PCA with the `imputePCA()` function. Note that here, the imputed values are stored in the object `.$completeObs` (here, `sensory` is used):

```
library(missMDA)

imputePCA(sensory, quali.sup=1:4, method="EM")$completeObs %>%
  dplyr::select(Judge, Product, Sour, Light)
```

[2] It is worth noticing that the individual differences could also be included by simply adding the Judge effect in the model.

```
## # A tibble: 99 x 4
##   Judge Product  Sour Light
##   <chr> <chr>   <dbl> <dbl>
## 1 J01   P01         0  22.8
## 2 J01   P02         0  21
## 3 J01   P03         0  20.4
## 4 J01   P04         0  33.7
## # ... with 95 more rows
```

In this case, it can be seen that the missing value for $J01 \times P04 \times Light$ has been replaced by the value `33.7`.

### *Limitations*

As we have seen, there are different ways to implement missing values, and the different algorithms will likely impute them with different values. Consequently, the overall results can be affected and there is no way to know which solution is the most suitable for our study. Still, it is recommended to treat the missing values and to chose the right strategy that is the most adapted to the data.

However, since most imputation methods involve modeling, applying them to variables with a high missing values rate can introduce bias in the data. Let's consider a situation in which assessors are evaluating half the product set using a BIB. This means that, by design, half of the data are missing. By imputing the missing values, each prediction is proportionally based on one unique value. And ultimately, any further analyses on this data would be based on half measured and half *fictive* data.

## 9.2.3 Design Inspection

The next point of interest – quite specific to sensory and consumer data – is to ensure that the design is well balanced, and correctly handles the first-order and carry-over effects. This step is particularly important for those who analyze the data but were not involved from the start in that study (and hence were not involved in the test set-up).

Let's show a simple procedure that would check part of the quality of a design. Since our data set stored in *biscuits_sensory_profile.xlsx* does not contain any information regarding the experimental design, let's use `sensochoc` from {`SensoMineR`} instead.

To load (and clean) the data, let's run these lines of code:

```
library(SensoMineR)
data(chocolates)
```

```
dataset <- sensochoc %>%
  as_tibble() %>%
  mutate(across(c(Panelist, Session, Rank, Product), as.character))
```

The data consist of 6 products (`Product`) evaluated by 29 panelists (`Panelist`) in duplicates (`Session`). The presentation order is stored in `Rank`.

To evaluate whether the products have been equally presented at each position, a simple cross-count between `Product` and `Rank` is done. This can be done using the `xtabs()` function:

```
xtabs(~Product + Rank, data=dataset)
```

```
##         Rank
## Product  1  2  3  4  5  6
##   choc1  9  9 10  9 11 10
##   choc2 11  9  9 11  7 11
##   choc3  9 11 10  9  9 10
##   choc4  9 10  9 10 10 10
##   choc5 11  8 11 10 10  8
##   choc6  9 11  9  9 11  9
```

Such a table can also be obtained using `group_by()` and `count()` to get the results in a tibble:

```
dataset %>%
  group_by(Product) %>%
  count(Rank) %>%
  ungroup() %>%
  pivot_wider(names_from=Rank, values_from=n)
```

As we can see, the design is not perfectly balanced, as `choc2` is evaluated 11 times in the 1st, 4th, and 6th position, but only 7 times in the 5th position.

To make sure that the design is well balanced in terms of carry-over effect, we need to count how often each product is tested before each of the other products. Since this information is not directly available in the data, it needs to be added.

Let's start with extracting the information available, i.e. the order of each product for each panelist and session:

```
current <- dataset %>%
  dplyr::select(Panelist, Product, Session, Rank) %>%
  mutate(Rank = as.numeric(Rank))
```

An easy way to add the `Previous` product information as a new column in the data is by replacing `Rank` by `Rank + 1` in `current` (all new positions larger than the number of products are filtered).

```
previous <- current %>%
  rename(Previous = Product) %>%
  mutate(Rank = Rank + 1) %>%
  filter(Rank <= length(unique(dataset$Product)))
```

These new data are merged to `current` by `Panelist`, `Session`, and `Rank`:

```
cur_prev <- current %>%
  left_join(previous, by=c("Panelist", "Session", "Rank"))
```

As can be seen, the products that are evaluated first get `NA` in `Previous`, and for each rank r (r > 1), `Previous` gets the product that was evaluated at rank r−1.

To evaluate whether the carry-over effect is well-balanced, the only thing left to do is cross-count `Product` and `Previous` (here, the results are split per `Session`):

```
cur_prev %>%
  group_by(Session, Product, Previous) %>%
  count() %>%
  ungroup() %>%
  mutate(Product = factor(Product,
                          levels=paste0("choc", 1:6)),
         Previous = factor(Previous,
                           levels=c("NA",paste0("choc", 1:6)))) %>%
  arrange(Previous) %>%
  pivot_wider(names_from=Previous, values_from=n, values_fill=0) %>%
  arrange(Product) %>%
  split(.$Session)
```

```
## $`1`
## # A tibble: 6 x 9
##   Session Product choc1 choc2 choc3 choc4 choc5 choc6
##   <chr>   <fct>   <int> <int> <int> <int> <int> <int>
## 1 1       choc1       0     5     6     4     5     5
## 2 1       choc2       3     0     5     6     4     5
## 3 1       choc3       5     4     0     5     6     5
## 4 1       choc4       5     5     5     0     4     5
## # ... with 2 more rows, and 1 more variable:
## #   `NA` <int>
##
## $`2`
## # A tibble: 6 x 9
##   Session Product choc1 choc2 choc3 choc4 choc5 choc6
##   <chr>   <fct>   <int> <int> <int> <int> <int> <int>
## 1 2       choc1       0     4     5     5     6     4
## 2 2       choc2       3     0     5     5     6     5
## 3 2       choc3       5     4     0     6     5     4
## 4 2       choc4       5     5     5     0     4     6
## # ... with 2 more rows, and 1 more variable:
## #   `NA` <int>
```

As expected, the table shows that a product is never evaluated twice in a row (the diagonal contains 0s). Here again, the design is not optimal since `choc1` has been evaluated three times before `choc2` and six times before `choc5` in the first session.

---

The last column defined as `NA` refers to the number of time that products did not have a product tested before. In other words, this indicates that they were evaluated first.

---

## 9.3 Clean

As mentioned in the introduction of this chapter, there is a thin line between *Data Inspection* and *Data Manipulation*, as both steps share many common features. Here, we are limiting ourselves on handling variables and their type. For a full overview, we encourage the readers to look at Chapter 4 to see other ways to handle data.

### 9.3.1 Handling Data Type

The data used in this section are stored in *bisuits_traits.xlsx*. So let's start with importing it in R:

```
file_path <- here("Data", "biscuits_traits.xlsx")

demo_var <- read_xlsx(file_path, sheet="Variables") %>%
  dplyr::select(Code, Name)

demo_lev <- read_xlsx(file_path, sheet="Levels") %>%
  dplyr::select(Question, Code, Levels) %>%
  inner_join(demo_var, by=c("Question"="Code")) %>%
  dplyr::select(-Question)

demographic <- read_xlsx(file_path, sheet="Data", skip=1,
                         col_names=unlist(demo_var$Name))
```

In R, the variables can be of different types, going from numerical to nominal to binary, etc. This section aims at presenting the most common types (and their properties) used in sensory and consumer studies and at showing how to transform a variable from one type to another.

Remember that when your data set is stored in a tibble (as is the case here), the type of each variable is provided as sub-header when printed on screen. This eases the work of the analyst as the variables' type can be accessed at any moment. In case the data are not in a tibble, the use of the `str()` function becomes handy as it provides this information (here we limit ourselves to the first five columns).

```
str(demographic[,1:5])
```

```
## tibble [107 x 5] (S3: tbl_df/tbl/data.frame)
##  $ Living area: num [1:107] 1 1 2 1 1 1 1 1 3 1 ...
##  $ Housing    : num [1:107] 1 1 2 2 1 1 1 1 2 2 ...
##  $ Judge      : chr [1:107] "J48" "J61" "J60" "J97" ...
##  $ Height     : num [1:107] 1.45 1.6 1.62 1.6 1.69 1.62 1.58 1.6 1.56 1.67 ...
##  $ Weight     : num [1:107] 43 65 52 60 70 56 62 55 55 53 ...
```

In sensory and consumer research, the four most common types are:

- Numerical (incl. integer [`int`], decimal [`dcl`], and double [`dbl`]);
- Logical [`lgl`];
- Character [`char`];
- Factor [`fct`].

R still has plenty of other types, for more information please visit: https://tibble.tidyverse.org/articles/types.html

### Numerical Data

Since a large proportion of sensory research is quantitative, it is no surprise that our data are often dominated with numerical variables. In practice, numerical data include integers (non-fractional number, e.g. 1, 2, −16, etc.) or decimal values (or double, e.g. 1.6, 2.333333, −3.2, etc.).

By default, when reading data from an external file, R converts any numerical variables to integer unless decimal points are detected, in which case it is converted into double.

### Binary Data

Another common type that seems to be numerical in appearance, but that has additional properties is the binary type. Binary data are data that take two possible values (`TRUE` or `FALSE`), and are often the results of a *test* (e.g. is `x>3`? Or is `MyVar` numerical?). A typical example of binary data in sensory and consumer research is data collected through Check-All-That-Apply (CATA) questionnaires.

---

Intrinsically, binary data are *numerical*, TRUE being assimilated to 1 and FALSE to 0. If multiple tests are being performed, it is possible to sum the number of tests that pass using the `sum()` function, as shown in the simple example below: `# Generate 10 random values between 1 and 10 (uniform distribution) x <- runif(10, 1, 10) # Test whether the values generated are strictly larger than 5 test <- x>5 # Counting the number of values strictly larger than 5 sum(test)`

---

### Nominal Data

Nominal data are any data that are defined through text or strings. Note that nominal data would occur when variables are defined with numbers, although they do not have a numerical meaning. This is for instance the case when the respondents or samples are identified through numerical codes. But since the software cannot guess that those numbers are *identifiers* rather than *numbers*, the variables should be declared as nominal. The procedure explaining how to convert the type of the variables is detailed in Section 9.3.2.

For nominal data, two particular types of data are of interest:

- Character or `char`;
- Factor or `fct`.

Variables defined as character or factor take strings as input. However, these two types differ in terms of structure of their levels:

- For character, there are no particular structures, and the variables can take any values (e.g. open-ended question);
- For factor, the inputs of the variables are structured into levels.

To evaluate the number of levels, different procedures are required:

- For character, one should count the number of unique elements using length() and unique();
- For factor, the levels and the number of levels are directly provided by levels() and nlevels().

Let's compare a variable set as factor and character by using a simple made-up example:

```
example <- demographic %>%
  dplyr::select(Judge) %>%
  mutate(Judge_fct = as.factor(Judge))

summary(example)
```

```
##     Judge              Judge_fct
##  Length:107         J1      :  1
##  Class :character   J10     :  1
##  Mode  :character   J100    :  1
##                     J101    :  1
##                     J103    :  1
##                     J105    :  1
##                     (Other):101
```

```
#unique(example$Judge)
length(unique(example$Judge))
```

```
## [1] 107
```

```
#levels(example$Judge_fct)
nlevels(example$Judge_fct)
```

```
## [1] 107
```

Although `Judge` and `Judge_fct` look the same, they are structurally different, and those differences play an important role that one should consider when running certain analyses or for building tables and graphs.

The number of levels of a variable is directly read from the data when set as `character`, and its levels' order matches the way they appear in the data (or sometimes are rearranged in alphabetical order). This means that any data collected using a structured scale will often lose their natural order.

When set as `factor`, the factor levels (including their order) are informed, and do not depend necessarily on the data itself: If a level has never been selected, or if certain groups have been filtered, this information is still present in the data. In our case, the levels are read from the data and are reordered alphabetically (note that `J10` and `J100` appear before `J2` for instance.)

To illustrate this, let's rearrange the levels from `Judge_fct` by ordering them numerically in such a way `J2` follows `J1` rather than `J10`.

```
example <- demographic %>%
  dplyr::select(Judge) %>%
  mutate(Judge_fct = factor(Judge, str_sort(Judge, numeric=TRUE)))
levels(example$Judge_fct)[1:10]
```

```
##  [1] "J1"  "J2"  "J3"  "J4"  "J5"  "J6"  "J7"  "J8"
##  [9] "J9"  "J10"
```

Now the levels are sorted. Let's then filter respondents by only keeping J1 to J20. We then rerun the previous code that counts the number of elements in each variables:

```
example_reduced <- example %>%
  filter(Judge %in% paste0("J",1:20))

# unique(example_reduced$Judge)
length(unique(example_reduced$Judge))
```

```
## [1] 19
```

```
# levels(example_reduced$Judge_fct)
nlevels(example_reduced$Judge_fct)
```

```
## [1] 107
```

After filtering some respondents, it can be noticed that the variable set as character only contains 19 elements (`J18` doesn't exist in the data), whereas the column set as factor still contains the 107 entries (most of them not having any recordings).

```
example_reduced %>%
  count(Judge, .drop=FALSE)
```

```
## # A tibble: 19 x 2
##   Judge      n
##   <chr> <int>
## 1 J1         1
## 2 J10        1
## 3 J11        1
## 4 J12        1
## # ... with 15 more rows
```

```
example_reduced %>%
  count(Judge_fct, .drop=FALSE)
```

```
## # A tibble: 107 x 2
##   Judge_fct     n
##   <fct>     <int>
## 1 J1            1
## 2 J2            1
## 3 J3            1
## 4 J4            1
## # ... with 103 more rows
```

This property can be seen as an advantage or a disadvantage depending on the situation:

- For frequencies, it may be relevant to remember all the options, including the ones that may never be selected and to order the results logically (use of `factor`).
- For hypothesis testing (e.g. ANOVA) on a subset of data, the `Judge` variable set as `character` would have the correct number of degrees of freedom (18 in our example), whereas the variable set as factor would still use the original count (so 106 here)!

The latter point is particularly critical since the analysis is incorrect and will either return an error or (worse!) return erroneous results!

Last but not least, variables defined as factor allow having their levels being renamed (and eventually combined) very easily. Let's consider the `Living area` variable from `demographic` as an example. From the original excel file, it can be seen that it has three levels, `1` corresponding to *urban area*, `2` to *rurban area*, and `3` to *rural area*. Let's start by renaming its levels:

```
example = demographic %>%
  mutate(Area = factor(`Living area`, levels=c(1,2,3),
                       labels=c("Urban", "Rurban", "Rural")))

levels(example$Area)
```

```
## [1] "Urban"  "Rurban" "Rural"
```

```
nlevels(example$Area)
```

```
## [1] 3
```

```
table(example$`Living area`, example$Area)
```

```
##
##     Urban Rurban Rural
##   1    72      0     0
##   2     0     12     0
##   3     0      0    23
```

As can be seen, the variable `Area` is the factor version (including its labels) of `Living area`. Let's now regroup `Rurban` and `Rural` together under `Rural`, and change the order to ensure that `Rural` appears before `Urban`:

```
example = demographic %>%
  mutate(Area = factor(`Living area`, levels=c(2,3,1),
                       labels=c("Rural", "Rural", "Urban")))

levels(example$Area)
```

```
## [1] "Rural" "Urban"
```

```
nlevels(example$Area)
```

```
## [1] 2
```

```
table(example$`Living area`, example$Area)
```

```
##
##      Rural Urban
##   1      0    72
##   2     12     0
##   3     23     0
```

This approach of renaming and reordering factor levels is very important as it can simplify the readability of tables and figures. Some other transformations can be applied to factors thanks to the {`forcats`} package. Particular attention should be given to the following functions:

- `fct_reorder()`/`fct_reorder2()` and `fct_relevel()` reorder the levels of a factor;
- `fct_recode()` renames the factor levels (as an alternative to `factor()` used in the previous example);
- `fct_collapse()` and `fct_lump()` aggregate different levels together (`fct_lump()` regroups automatically all the rare levels);
- `fct_inorder()` uses the order read in the data (particularly useful with `pivot_longer()` for instance);
- `fct_rev()` reverses the order of the levels (particularly useful in graphs).

Although it has not been done here, manipulating strings is also possible through the {`stringr`} package, which provides interesting functions such as:

- `str_to_upper()`/`str_to_lower()` to convert strings to uppercase or lowercase;
- `str_c()` to combine and `str_sub()` to subset strings;
- `str_trim()` and `str_squish()` remove white spaces;
- `str_extract()`, `str_replace()`, or `str_split()` extract, replace, or split strings or part of the strings;
- `str_sort()` to order alphabetically (or by respecting numbers, as shown previously) its elements.

Many of these functions will be used later in Chapter 13.

### 9.3.2 Converting between Types

Since each variable type has its own properties, it is important to be able to switch from one to another if needed. This can be critical (converting from numerical to character or factor and reversely) or purely practical (converting from character to factor and reversely).

In the previous section, we have already seen how to convert from character to factor. Let's now consider two other conversions, namely:

- from numerical to character/factor;
- from character/factor to numerical.

The conversion from numerical to character or factor is simply done using `as.character()` and `as.factor()`, respectively. An example in the use of `as.character()` and `as.factor()` was provided in the previous section when we converted the `Respondent` variables to character and factor. The use of `factor()` was also used earlier when the variable `Living area` was converted from numerical to factor (called `Area`) with labels.

---

`as.factor()` only converts into factors without allowing to chose the order of the levels nor to rename them. Instead, `factor()` should be used as it allows specifying the `levels` (and hence the order of the levels) and their corresponding `labels`.

---

To illustrate the conversion from character to numeric, let's start with creating a tibble with two variables, one containing strings made of numbers and another one containing strings made of text.

```
example <- tibble(Numbers = c("2","4","9","6","8","12","10"),
                  Text = c("Data","Science","4","Sensory",
                           "and","Consumer","Research"))
```

The conversion from character to numerical is straightforward and requires the use of the function `as.numeric()`:

```
example %>%
  mutate(NumbersN = as.numeric(Numbers), TextN = as.numeric(Text))
```

```
## Warning: There was 1 warning in 'mutate()'.
## i In argument: 'TextN = as.numeric(Text)'.
## Caused by warning:
```

```
## ! NAs introduced by coercion

## # A tibble: 7 x 4
##   Numbers Text    NumbersN TextN
##   <chr>   <chr>      <dbl> <dbl>
## 1 2       Data           2    NA
## 2 4       Science        4    NA
## 3 9       4              9     4
## 4 6       Sensory        6    NA
## # ... with 3 more rows
```

As can be seen, when strings are made of numbers, the conversion works fine. However, any non-numerical string character cannot be converted and hence returns NAs.

Now let's apply the same principle to a variable of the type factor. To do so, the same example in which the variables are now defined as factor is used:

```
example <- example %>%
  mutate(Numbers = as.factor(Numbers)) %>%
  mutate(Text = factor(Text, levels=c("Data","Science","4","Sensory",
                                      "and","Consumer","Research")))
```

Let's apply as.numeric() to these variables:

```
example %>%
  mutate(NumbersN = as.numeric(Numbers), TextN = as.numeric(Text))
```

```
## # A tibble: 7 x 4
##   Numbers Text    NumbersN TextN
##   <fct>   <fct>      <dbl> <dbl>
## 1 2       Data           3     1
## 2 4       Science        4     2
## 3 9       4              7     3
## 4 6       Sensory        5     4
## # ... with 3 more rows
```

We can notice here that the outcome is not really what was expected. The numbers 2-4-9-6-8-12-10 become 3-4-7-5-6-2-1, and Data-Science-4-Sensory-and-Consumer-Research becomes 1-2-3-4-5-6-7. The rationale behind this conversion is that the numbers do not reflect the string itself, but the position of that level in the factor level order.

To properly convert numerical factor levels to number, the variable should first be converted into character:

```
example %>%
  mutate(Numbers = as.numeric(as.character(Numbers)))
```

```
## # A tibble: 7 x 2
##   Numbers Text
##     <dbl> <fct>
## 1       2 Data
## 2       4 Science
## 3       9 4
## 4       6 Sensory
## # ... with 3 more rows
```

As can be seen, it is very important to verify the type of each variable (and convert it if needed) to ensure that the data are processed as they should be. Since each type has its own advantages and drawbacks, it is convenient to regularly switch from one to another. Don't worry, you will quickly familiarize with this as we will be doing such conversions regularly in the next sections.

# 10

# *Data Analysis*

Although the data-science workflow suggests a clear separation between data manipulation and data analysis, in practice such separation is not that obvious. Indeed, most analyses require data manipulation. In fact, some data transformation can be seen as a part of both data transformation and data analysis. Yet, this section is somewhat more dedicated to the analysis of data by 1) presenting how some of the most common analyses in sensory and consumer science are performed, 2) integrating the analysis part to your script, and most importantly 3) providing applications and extensions (or alternatives) to all the procedures presented in Chapters 4 and 5. With that in mind, the emphasis is not on the results and interpretation of the results, but on the path to get such results. For practical reasons, this chapter is divided in three subsections, one dedicated to the sensory data, one to the consumer data, and one combining both.

## 10.1 Sensory Data

As one may expect, this chapter is mostly built around the {tidyverse}:

```
library(tidyverse)
library(here)
library(readxl)
```

Let's start with the analysis of our sensory data stored in *biscuits_sensory_profile.xlsx*.

```
file_path <- here("data", "biscuits_sensory_profile.xlsx")
p_info <- readxl::read_xlsx(file_path, sheet = "Product Info") %>%
```

DOI: 10.1201/9781003028611-10

```
  dplyr::select(-Type)

sensory <- readxl::read_xlsx(file_path, sheet = "Data") %>%
  inner_join(p_info, by = "Product") %>%
  relocate(Protein:Fiber, .after = Product)
```

Typically, sensory scientists first seek to determine whether there are differences between samples for the different attributes. This is done through Analysis of Variance (ANOVA) and can be achieved using the `lm()` or `aov()` functions.

Let's start by running the ANOVA for the attribute `Sweet`. Since the test has not been duplicated, a two-way ANOVA (including the Product and Assessor effects) without interaction is used. This is done using the following code:

```
sweet_aov <- lm(Sweet ~ Product + Judge, data = sensory)
anova(sweet_aov)
```

```
## Analysis of Variance Table
##
## Response: Sweet
##           Df Sum Sq Mean Sq F value  Pr(>F)
## Product   10   2654     265    7.27 4.5e-08 ***
## Judge      8   4451     556   15.25 2.3e-13 ***
## Residuals 80   2918      36
## ---
## Signif. codes:
## 0 '***' 0.001 '**' 0.01 '*' 0.05 '.' 0.1 ' ' 1
```

---

The results provided here by `anova()` are not very convenient as the output is not stored in a matrix or a data frame. We will illustrate later how to apply the `tidy()` function from {`broom`} to tidy the statistical output from most testing/modelling functions into a user-friendly tibble.

---

We could duplicate this code for each single attribute, but this would be quite tedious for a large number of attributes. Moreover, this code is sensitive to the way the variables are named and hence might not be suitable for other data sets. Instead, we propose two solutions, one using `split()` in combination with `map()` and one involving `nest_by()` to run this analysis automatically.

For both these solutions, the data should be stored in the long and thin form, which can be obtained using `pivot_longer()`:

```
senso_aov_data <- sensory %>%
  pivot_longer(Shiny:Melting,
               names_to = "Attribute", values_to = "Score")
```

From this structure, the first approach consists of splitting the data by attribute. Once done, we run the ANOVA for each subset (the model is then defined as `Score ~ Product + Judge`) automatically using `map()`,[1] and we extract the results of interest using the {`broom`} package.

Ultimately, the results can be combined again using `enframe()` and `unnest()`.

```
senso_aov1 <- senso_aov_data %>%
  split(.$Attribute) %>%
  map(function(data) {
    res <- broom::tidy(anova(lm(Score ~ Product + Judge, data = data)))
    return(res)
  }) %>%
  enframe(name = "Attribute", value = "res") %>%
  unnest(res)
```

The second approach uses the advantage of tibbles and nests the analysis by attribute (meaning the analysis is done for each attribute separately, a bit like `group_by()`). In this case, we store the results of the ANOVA in a new variable called `mod`.

Once the analysis is done, we extract the info stored in `mod` by converting it into a tibble using {`broom`} and restructure it using `reframe()`:

```
senso_aov2 <- senso_aov_data %>%
  nest_by(Attribute) %>%
  mutate(mod = list(lm(Score ~ Product + Judge, data = data))) %>%
  reframe(broom::tidy(anova(mod))) %>%
  ungroup()
```

The two approaches return the exact same results:

```
## # A tibble: 96 x 7
##   Attribute  term     df sumsq meansq stati~1   p.value
##   <chr>      <chr> <int> <dbl>  <dbl>   <dbl>     <dbl>
## 1 Astringent Prod~    10  870.   87.0    1.62  1.16e- 1
```

[1] The `map()` function applies the same function to each element of a list automatically: It is hence equivalent to a `for()` loop, but in a neater and more efficient way.

```
## 2 Astringent Judge      8 5041.  630.      11.7   6.94e-11
## 3 Astringent Resi~     80 4302.    53.8    NA     NA
## 4 Bitter     Prod~     10 1005.  101.       3.95  2.11e- 4
## # ... with 92 more rows, and abbreviated variable name
## #   1: statistic
```

Let's dig into the results by extracting the attributes that do not show significant differences at 5%. Since the `tidy()` function from {`broom`} tidies the data into a tibble, all the usual data transformation can be performed. Let's filter only the `Product` effect under `term`, and let's order the `p.value` decreasingly:

```
res_aov <- senso_aov1 %>%
  filter(term == "Product") %>%
  dplyr::select(Attribute, statistic, p.value) %>%
  arrange(desc(p.value)) %>%
  mutate(p.value = round(p.value, 3))
```

```
res_aov %>%
  filter(p.value >= 0.05)
```

```
## # A tibble: 4 x 3
##   Attribute      statistic p.value
##   <chr>              <dbl>   <dbl>
## 1 Cereal flavor       1.22   0.294
## 2 Roasted odor        1.40   0.193
## 3 Astringent          1.62   0.116
## 4 Sticky              1.67   0.101
```

As can be seen, the products do not show any significant differences at 5% for 4 attributes: `Cereal flavor` (p=0.294), `Roasted odor` (p=0.193), `Astringent` (p=0.116), and `Sticky` (p=0.101).

Rather than showing the results in a table, let's visualize them graphically as a bar chart by representing the F-values (Figure 10.1). The attributes are ordered decreasingly and color-coded based on their significance:

```
res_aov %>%
  mutate(Signif = ifelse(p.value <= 0.05, "Signif.", "Not Signif.")) %>%
  mutate(Signif = factor(Signif,
                         levels = c("Signif.", "Not Signif."))) %>%
  ggplot(aes(x=reorder(Attribute, statistic), y=statistic,
             fill=Signif)) +
  geom_bar(stat = "identity") +
```

```
scale_fill_manual(values = c("Signif." = "forestgreen",
                             "Not Signif." = "orangered2")) +
ggtitle("Sensory Attributes",
        "(The attributes are sorted based on F-values (product))") +
theme_bw() +
xlab("") +
ylab("F-values") +
coord_flip()
```

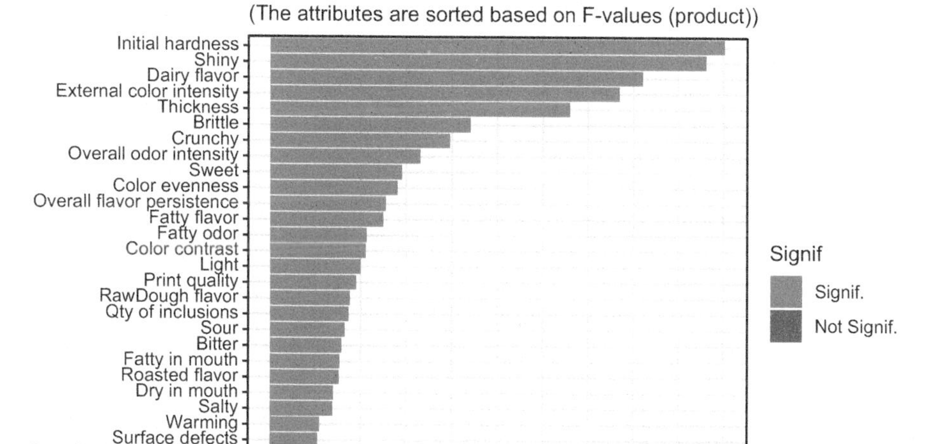


**FIGURE 10.1**
Sensory attributes ranked by decreasing F-values of the product effect in their respective ANOVA.

The evaluated biscuits appear to differ the most (top five attributes) for `Initial hardness`, `Shiny`, `Dairy flavor`, `External color intensity`, and `Thickness`.

---

As an alternative, the `decat()` function from the {`SensoMineR`} package would do the same job as it automatically performs ANOVAs on a set of attributes (presented in subsequent columns). Additionally, it also performs some t-tests that highlight which samples are significantly more (or less) intense than the average for each attribute.

---

Once the significant differences have been checked, a follow-up analysis consists of visualizing these differences in a multivariate way. Such visualization is often done through Principal Component Analysis (PCA). In practice, PCA is performed on the mean sensory profiles. Let's start with building such a table:

```
senso_mean <- sensory %>%
  pivot_longer(Shiny:Melting,
               names_to = "Attribute", values_to = "Score") %>%
  dplyr::select(-Judge) %>%
  pivot_wider(names_from = Attribute, values_from = Score,
              values_fn = mean)
```

The resulting table is then submitted to PCA. R proposes many solutions to run such analysis, including the `prcomp()` and `princomp()` functions from the {`stats`} package. However, we prefer to use `PCA()` from the {`FactoMineR`} as it is more complete and proposes many options that are very useful in sensory and consumer science (e.g. it generates the graphics automatically and allows projecting supplementary individuals and/or variables).

It should, however, be noted that the `PCA()` function does not accept tibbles. Instead, the table should be stored in a matrix or data frame format that contains the observation names (here the product names) as row names. Fortunately, a tibble can be easily converted into a data frame (`as.data.frame()`) and the `Product` column can be moved to row names (`column_to_rownames(var="Product")`).

Note that these data also contain two qualitative variables, `Protein` and `Fiber`. They should thus either be removed prior to running the analysis or (better!) be projected as supplementary variables through the `quali.sup` parameter from `PCA()`. Finally, since `POpt` is an optimized product whose sensory description was later added to the data set, let's not include it in the analysis per se (it is not contributing to the construction of the dimensions). Instead, we recommend to project it as a supplementary observation (through `ind.sup`) to illustrate where it would be located in the initial space.

```
library(FactoMineR)

senso_pca <- senso_mean %>%
  arrange(Product) %>%
  as.data.frame() %>%
  column_to_rownames(var = "Product") %>%
  PCA(., ind.sup = nrow(.), quali.sup = 1:2, graph = FALSE)
```

Since we set the option `graph=FALSE`, the PCA plots are not yet being generated. Although `PCA()` can generate the plots either in {`base`} R language or in {ggplot2}, we prefer to use a complementary package called {factoextra} which recreates most plots from {FactoMineR} (and some other packages) as a ggplot() object. This comes in very handy as you can benefit from the flexibility offered by ggplot().

The score plot (i.e. the product map) from PCA() is created through fviz_pca_ind(), whereas the representation of variable loadings is created with fviz_pca_var(). Eventually, fviz_pca_biplot() is used to produce the so-called *biplot*.

To illustrate this, let's display the product map by coloring the products using the supplementary variables (Protein and Fiber content) (Figure 10.2). This can easily be done through the habillage parameter from fviz_pca_ind(), which can either take the name of the qualitative variable (e.g. "Protein") or a numerical value indicating its position (e.g. 2).

```
library(factoextra)

fviz_pca_ind(senso_pca, habillage = "Protein", repel = TRUE, mean.point = FALSE)
fviz_pca_ind(senso_pca, habillage = 2, repel = TRUE, mean.point = FALSE,
             palette = c("#FF8C00", "#949494"))
fviz_pca_var(senso_pca)
fviz_pca_biplot(senso_pca)
```

Here, repel=TRUE uses geom_text_repel() from {ggrepel} (rather than geom_text() from {ggplot2}) to avoid having labels overlapping.

On the first dimension, P10 is opposed to P09 and P03 as it is more intense for attributes such as Sweet and Dairy flavor, for example, and less intense for attributes such as Dry in mouth and External color intensity. On the second dimension, P08and P06 are opposed to P02 and P07 as they score higher for Qty of inclusions, and Initial hardness, and score lower for RawDough flavor and Shiny. POpt is located between P05 and P06.

---

Many more visualizations can be produced. Among others, let's mention:
* Scree plot showing the evolution of the eigenvalues across dimensions to help decide how many dimensions to consider;
* The representation of the product space on other dimensions (by default, dimension 1 and dimension 2 are shown);
* Representations of the (product or attribute) space in which the contribution or quality of representation of the elements are showcased.

---

For more information regarding the various options offered by {factoextra}, see Kassambara (2017b).

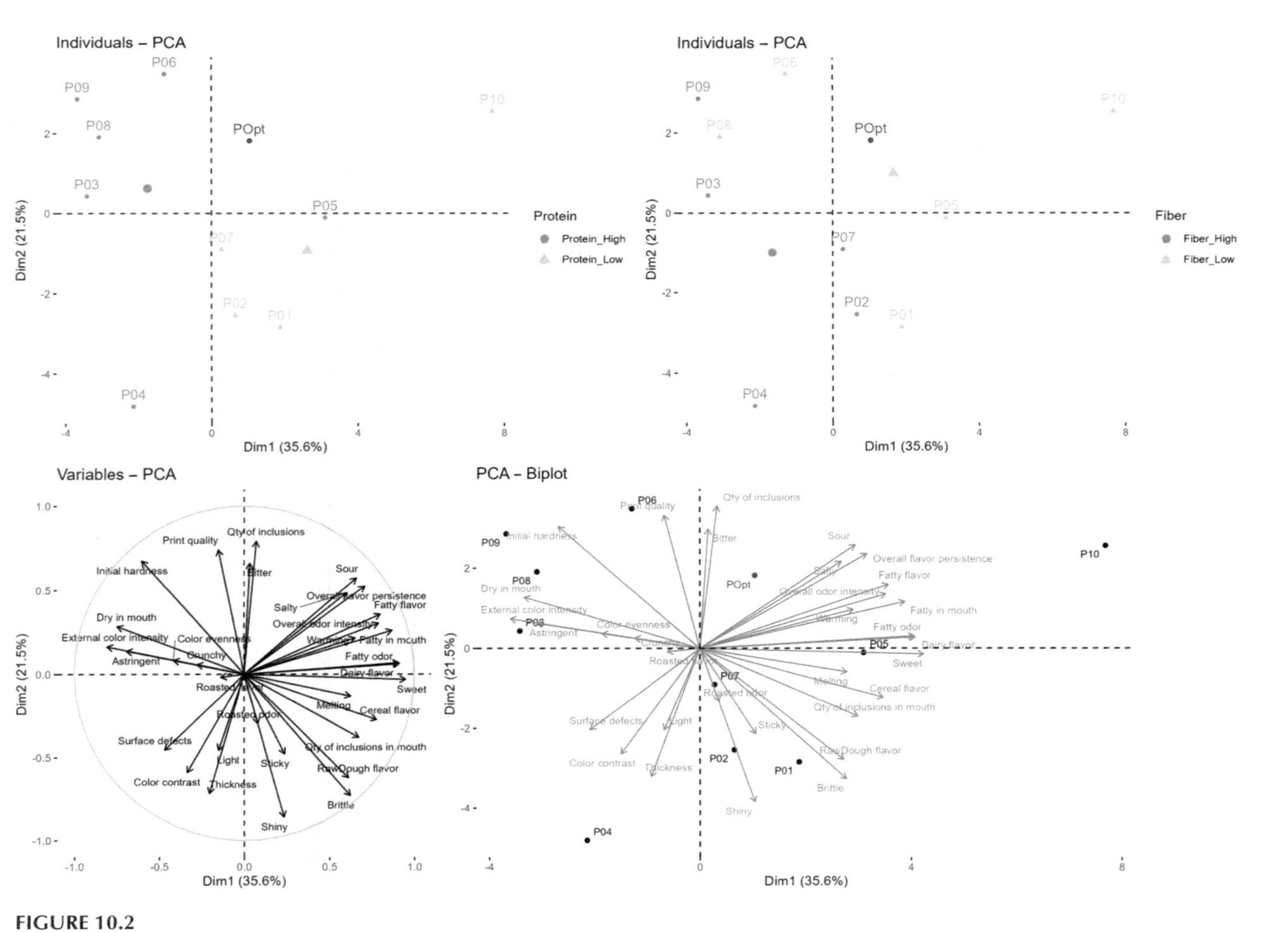


**FIGURE 10.2**
Display of PCA plots using *factoextra*.

## 10.2 Demographic and Questionnaire Data

The *biscuits_traits.xlsx* file contains descriptive (i.e. *demographic*) information regarding the consumers and their food-related behavioral traits (i.e. psychometric *TFEQ* data, see Chapter 7 for more information). This file has three tabs denoted as *Data*, *Variables*, and *Levels*:

- *Data* contains the data, which is coded;
- *Variables* provides information (e.g. name, information) related to the different variables present in *Data*;
- *Levels* provides information about the different levels each variable can take.

Let's start with importing this data set. The importation is done in multiple steps as follows:

```
file_path <- here("Data", "biscuits_traits.xlsx")
excel_sheets(file_path)
```

```
## [1] "Data"      "Variables" "Levels"
```

```
demo_var <- read_xlsx(file_path, sheet = "Variables") %>%
  dplyr::select(Code, Name)

demo_lev <- read_xlsx(file_path, sheet = "Levels") %>%
  dplyr::select(Question, Code, Levels) %>%
  inner_join(demo_var, by = c("Question" = "Code")) %>%
  dplyr::select(-Question)

demographic <- read_xlsx(file_path, sheet = "Data", skip = 1,
                         col_names = unlist(demo_var$Name))
```

### 10.2.1 Demographic Data: Frequency and Proportion

For this demographic data file, let's start by having a look at the partition of consumers for each of the descriptive variables. This is done by computing the frequency and proportion (in percentage) attached to each level of `Living area`, `Housing`, `Income range`, and `Occupation`. To obtain such a table, let's start by selecting only the columns corresponding to these variables together with `Judge`.

Since data from surveys and questionnaires are often coded (here, answer *#6* to question `Q10` means *Student*, while answer *#7* to the same question means *Qualified worker*), they first need to be decoded. In our case, the key to decode the data is stored in `demo_lev`.

Different strategies can be used to decode the data. One straight-forward strategy consists in automatically decoding each variable using `mutate()` and `factor()`. Another approach is considered here: Let's start with building a long thin tibble with `pivot_longer()` that we merge to `demo_lev` by `Question` and `Response` using `inner_join()`. We prefer this solution here as it is simpler, faster, and independent of the number of variables to decode.

Once done, we can aggregate the results by `Question` and `Levels` (since we want to use the level information, not their code) and compute the frequency (`n()`) and the proportion (`N/sum(N)`).[2]

```
library(formattable)

demog_reduced <- demographic %>%
  dplyr::select(Judge, `Living area`, Housing,
                `Income range`, `Occupation`) %>%
  pivot_longer(-Judge,
               names_to = "Question", values_to = "Response") %>%
  inner_join(demo_lev, by = c("Question" = "Name",
                              "Response" = "Code")) %>%
  group_by(Question, Levels) %>%
  summarize(N = n()) %>%
  mutate(Pct = percent(N / sum(N), digits = 1L)) %>%
  ungroup()
```

Bar plots and histograms are a nice way to visualize proportions and to compare them over several variables. Such plots can be obtained by splitting `demog_reduced` by `Question` and by creating them using either `N` or `Pct` (we are using `Pct` here). For simplicity, let's order the levels decreasingly (`reorder`) and represent them horizontally (`coord_flip()`). Of course, such graphs are automated across all questions using `map()` (Figures 10.3 to 10.6):

```
demog_reduced %>%
  split(.$Question) %>%
  map(function(data) {
    var <- data %>%
      pull(Question) %>%
      unique()
```

[2] We use the package {`formattable`} to print the results in percentage using one decimal. As an alternative, we could have used `percent()` from the {`scales`} package.

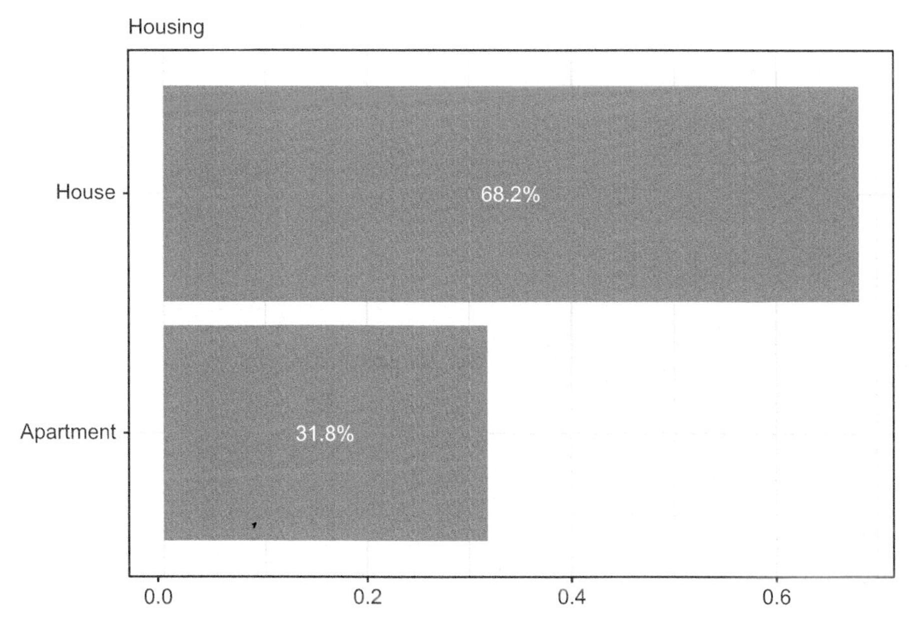


**FIGURE 10.3**
Bar plots showing the respondent distributions for the various demographic variables.

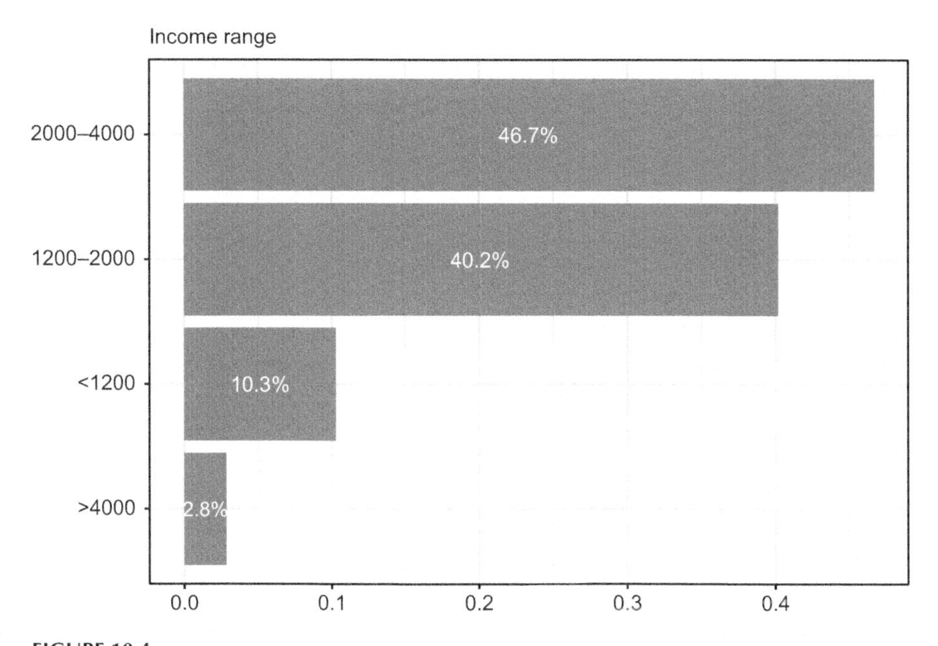


**FIGURE 10.4**
Bar plots showing the respondent distributions for the various demographic variables.

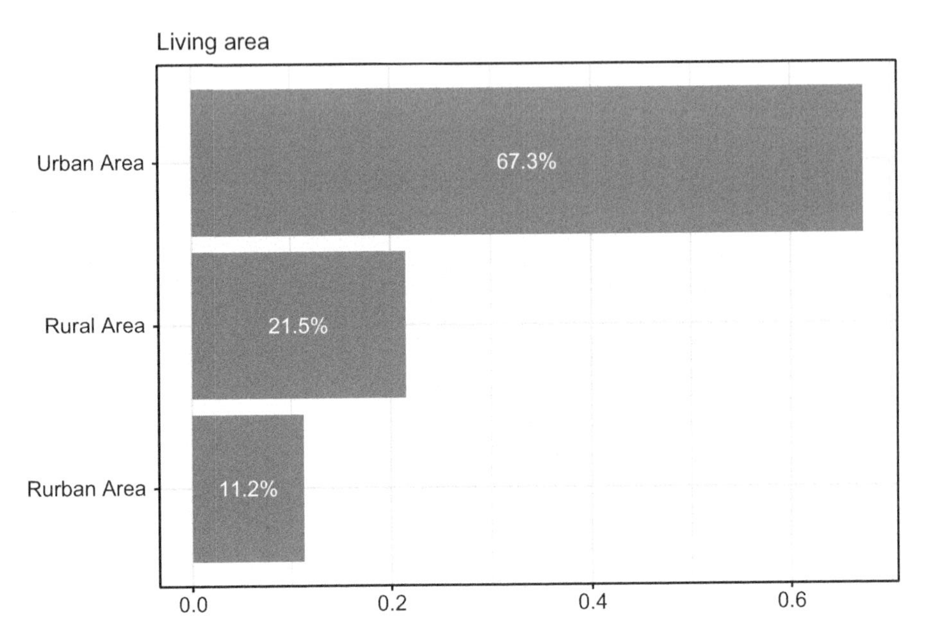


**FIGURE 10.5**
Bar plots showing the respondent distributions for the various demographic variables.

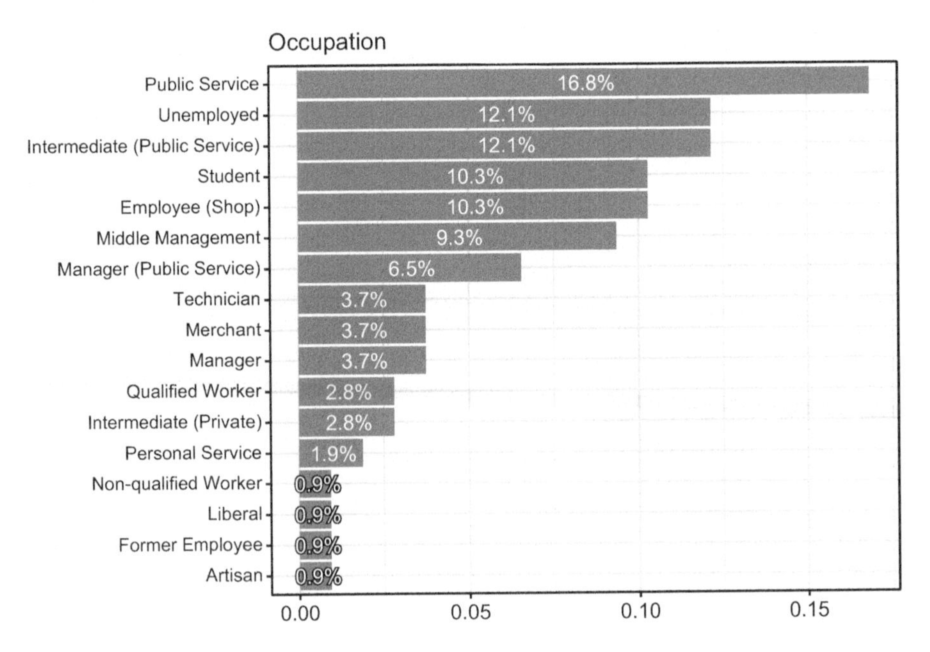


**FIGURE 10.6**
Bar plots showing the respondent distributions for the various demographic variables.

```
    ggplot(data, aes(x = reorder(Levels, Pct), y = Pct, label = Pct)) +
      geom_bar(stat = "identity", fill = "grey50") +
      geom_text(aes(y = Pct / 2), colour = "white") +
      xlab("") +
      ylab("") +
      ggtitle(var) +
      theme_bw() +
      coord_flip()
  })
```

### 10.2.2 Eating Behavior Traits: TFEQ Data

In the same data set, consumers also answered some questions that reflect their relation to food (Stunkard and Messick, 1985). These questions can be categorized into three groups (also known as factors):

- Disinhibition (variables starting with `D`);
- Restriction (variables starting with `R`);
- Sensitivity to Hunger (variables starting with `H`).

In order to analyze these three factors separately, we first need to select the corresponding variables. As we have seen earlier, such selection could be done by combining `dplyr::select()` to `starts_with("D")`, `starts_with("R")`, and/or `starts_with("H")`. However, this solution is not satisfactory as it also selects other variables that would start with any of these letters (e.g. Housing).

Instead, let's take advantage of the fact that variable names have a recurring pattern (they all start with the letters D, R, or H, followed by a number) to introduce the notion of *regular expressions*.

Regular expressions are coded expressions that allow finding patterns in names. In practice, generating a regular expression can be quite complex as it is an abstract concept which follows very specific rules. Fortunately, the package {`RVerbalExpression`} is a great assistant as it generates the regular expression for you thanks to understandable functions. To create a regular expression using {`RVerbalExpression`}, we should first initiate it by calling the function `rx()` to which any relevant rules can be added. In our case, the variables must start with any of the letter R, D, or H, followed by a number (or more, as values go from 1 to 21). This can be done using the following code:

```
library(RVerbalExpressions)
```

```
rdh <- rx() %>%
  rx_either_of(c("R", "D", "H")) %>%
  rx_digit() %>%
  rx_one_or_more()
```

`rdh` is defined as (R|D|H)+ which corresponds to the regular expression we were looking for. We can then reduce (through `dplyr::select()`) the table to the variables that fit our regular expression by using the function `matches()`.

```
demographic %>%
  dplyr::select(matches(rdh))
```

For each variable, let's create a frequency table. Although we could use already built-in functions, let's customize our table (including raw frequency and percentages) as we want by creating our own function (called here `myfreq()`):

```
myfreq <- function(data, info) {
  var <- unique(unlist(data$TFEQ))
  info <- info %>%
    filter(Name == var)

  res <- data %>%
    mutate(Response = factor(Response, levels = info$Code,
                             labels = info$Levels)) %>%
    arrange(Response) %>%
    group_by(Response) %>%
    summarize(N = n()) %>%
    mutate(Pct = percent(N / sum(N), digits = 1L)) %>%
    ungroup()

  return(res)
}
```

We then apply this function to each variable separately using `map()` after pivoting all these variables of interest (`pivot_longer()`) and splitting the data by `TFEQ` question:

```
TFEQ_freq <- demographic %>%
  dplyr::select(Judge, matches(rdh)) %>%
  pivot_longer(-Judge, names_to = "TFEQ", values_to = "Response") %>%
  split(.$TFEQ) %>%
  map(myfreq, info = demo_lev) %>%
  enframe(name = "TFEQ", value = "res") %>%
  unnest(res) %>%
  mutate(TFEQ = factor(TFEQ, levels = unique(
    str_sort(.$TFEQ, numeric=TRUE)))) %>%
  arrange(TFEQ)
```

From this table, histograms representing the frequency distribution for each variable can be created. But let's suppose that we only want to display variables related to Disinhibition (Figure 10.7). To do so, we first need to generate the corresponding regular expression (only selecting variables starting with "D") to filter the results before creating the plots:

```
d <- rx() %>%
  rx_find("D") %>%
  rx_digit() %>%
  rx_one_or_more()

TFEQ_freq %>%
  filter(str_detect(TFEQ, d)) %>%
  ggplot(aes(x = Response, y = Pct, label = Pct)) +
  geom_bar(stat = "identity", fill = "grey50") +
  geom_text(aes(y = Pct / 2), colour = "white") +
  theme_bw() +
  theme(axis.text = element_text(hjust = 1, angle = 30)) +
  facet_wrap(~TFEQ, scales = "free")
```

Structured questionnaires such as the TFEQ are very frequent in sensory and consumer science. They are used to measure individual patterns as diverse as personality traits, attitudes, food choice motives, engagement, social desirability bias, etc. Ultimately, the TFEQ questionnaire consists of a set of structured questions whose respective answers combine to provide a TFEQ score (actually, three scores, one for Disinhibition, one for Restriction, and one for sensitivity to Hunger). These TFEQ scores translate into certain food behavior tendencies.

However, computing the TFEQ scores is slightly more complicated than adding the scores of all TFEQ questions together. Instead, they follow certain rules that are stored in the *Variables* spreadsheet in *biscuits_traits.xlsx*. For each TFEQ question, the rule to follow is provided by `Direction` and `Value` and works as follows: if the condition provided by `Direction` and `Value` is

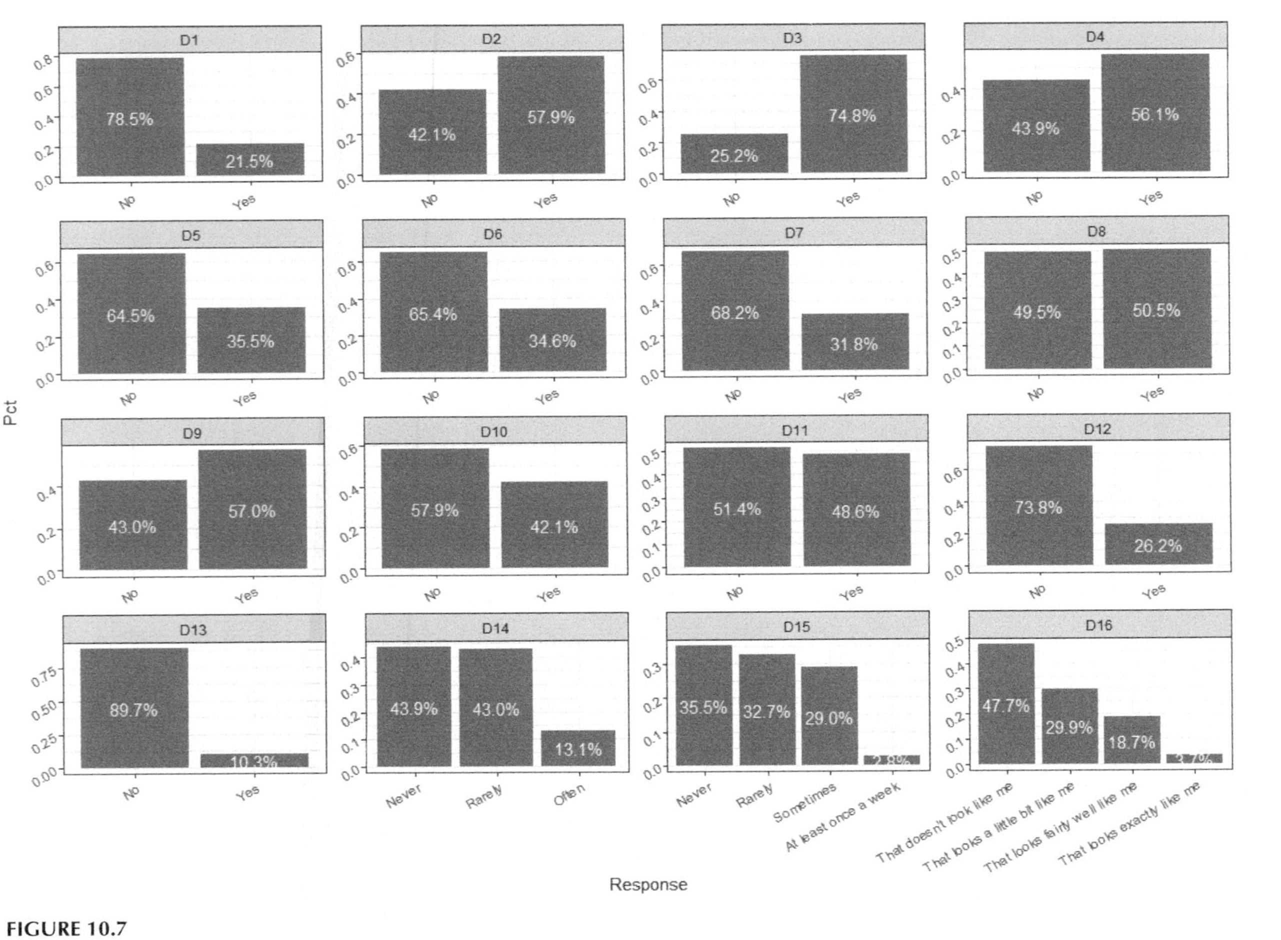


**FIGURE 10.7**
Frequency distributions for disinhibition-related variables only.

met, then the respondent gets a *1*, else a *0*. Ultimately, the TFEQ score is the sum of all these evaluations.

Let's start by extracting this information (`Direction` and `Value`) from the *Variables* sheet for all the variables involved in the computation of the TFEQ scores. We store this in `var_drh`.

```
var_rdh <- read_xlsx(file_path, sheet = "Variables") %>%
  filter(str_detect(Name, rdh)) %>%
  dplyr::select(Name, Direction, Value)
```

This information is added to `demographic`.

```
TFEQ <- demographic %>%
  dplyr::select(Judge, matches(rdh)) %>%
  pivot_longer(-Judge, names_to = "DHR", values_to = "Score") %>%
  inner_join(var_rdh, by = c("DHR" = "Name"))
```

Since we need to evaluate each assessors' answer to the TFEQ questions, we create a new variable `TFEQValue` which takes a 1 if the corresponding condition is met or a 0 otherwise. Such approach is done through `mutate()` combined with a succession of intertwined `ifelse()` functions.[3]

```
TFEQ_coded <- TFEQ %>%
  mutate(TFEQValue = ifelse(Direction == "Equal" & Score == Value, 1,
    ifelse(Direction == "Superior" & Score > Value, 1,
      ifelse(Direction == "Inferior" & Score < Value, 1, 0)
    )
  )) %>%
  mutate(Factor = ifelse(str_starts(.$DHR, "D"), "Disinhibition",
    ifelse(str_starts(.$DHR, "H"), "Hunger", "Restriction")
  )) %>%
  mutate(Factor = factor(Factor, levels = c("Restriction",
                                            "Disinhibition",
                                            "Hunger")))
```

Ultimately, we compute the TFEQ score by summing across all `TFEQValue` per respondent, by maintaining the distinction between each category. Note that the final score is stored in `Total`, which corresponds to sum across categories:

---

[3] The function `ifelse()` takes three parameters: 1. the condition to test, 2. the value or code to run if the condition is met, and 3. the value or code to run if the condition is not met.

```
TFEQ_score <- TFEQ_coded %>%
  group_by(Judge, Factor) %>%
  summarize(TFEQ = sum(TFEQValue)) %>%
  mutate(Judge = factor(Judge, levels = unique(
    str_sort(.$Judge, numeric = TRUE)))) %>%
  arrange(Judge) %>%
  pivot_wider(names_from = Factor, values_from = TFEQ) %>%
  mutate(Total = sum(across(where(is.numeric))))
```

```
## 'summarise()' has grouped output by 'Judge'. You can
## override using the '.groups' argument.
```

Such results can then be visualized graphically, for instance by representing the distribution of `TFEQ_score` for the three TFEQ factors (Figure 10.8):

```
TFEQ_score %>%
  dplyr::select(-Total) %>%
  pivot_longer(-Judge, names_to = "Factor", values_to = "Scores") %>%
  ggplot(aes(x = Scores, color = Factor)) +
```

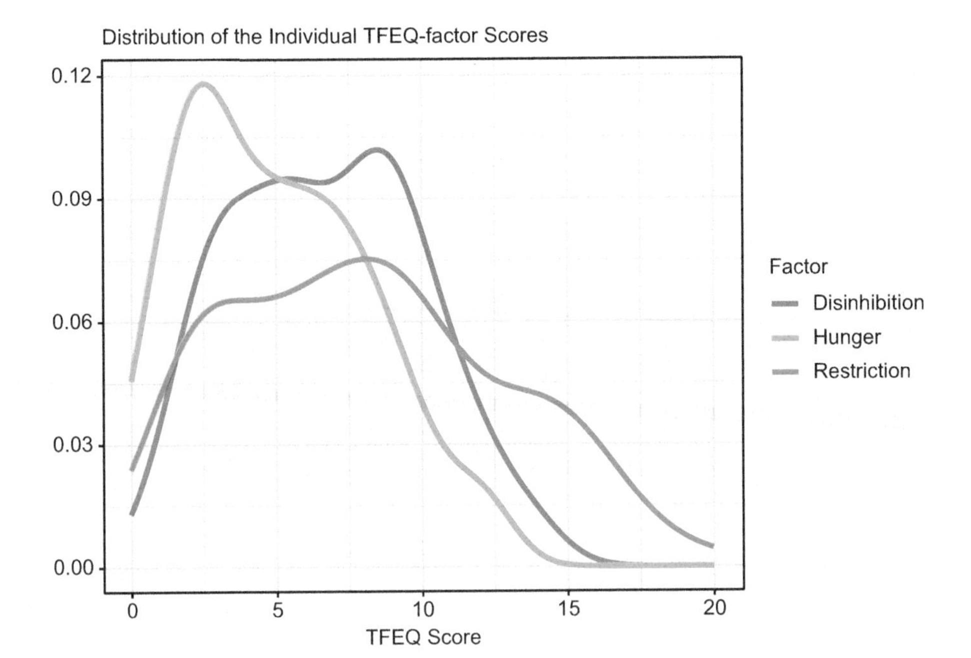


**FIGURE 10.8**
Visualization of the distributions of the TFEQ factors for the consumer panel.

```
geom_density(lwd = 1.5, key_glyph = "path") +
xlab("TFEQ Score") +
ylab("") +
guides(color = guide_legend(override.aes = list(linetype = 1))) +
ggtitle("Distribution of the Individual TFEQ-factor Scores") +
theme_bw()
```

## 10.3 Consumer Data

The analysis of consumer data usually involves the same type of analysis as the ones for sensory data (e.g. ANOVA, PCA, etc.), but the way the data are being collected (absence of repetitions) and their underlying nature (affect vs. descriptive) require some adjustments.

Let's start by importing the consumer data that are stored in *biscuits_consumer_test.xlsx*. Here, we import two spreadsheets, one with the consumption time and number of biscuits (stored in `Nbiscuit`) and one with different consumer evaluations of the samples (stored in `consumer`).

```
file_path <- here("Data", "biscuits_consumer_test.xlsx")

Nbiscuit <- read_xlsx(file_path, sheet = "Time Consumption") %>%
  mutate(Product = str_c("P", Product)) %>%
  rename(N = `Nb biscuits`)

consumer <- read_xlsx(file_path, sheet = "Biscuits") %>%
  rename(Judge = Consumer, Product = Samples) %>%
  mutate(Judge = str_c("J", Judge), Product = str_c("P", Product)) %>%
  inner_join(Nbiscuit, by = c("Judge", "Product"))
```

Similar to what we did with the sensory data, let's start with computing the mean liking scores per product after the first bite (`1stbite_liking`) and at the end of the evaluation (`after_liking`).

```
consumer %>%
  dplyr::select(Judge, Product, `1stbite_liking`, `after_liking`) %>%
  group_by(Product) %>%
  summarise(across(where(is.numeric), mean))
```

```
## # A tibble: 10 x 3
```

```
##   Product '1stbite_liking' after_liking
##   <chr>                <dbl>        <dbl>
## 1 P1                    6.30         6.26
## 2 P10                   7.40         7.57
## 3 P2                    5.53         5.38
## 4 P3                    3.94         3.49
## # ... with 6 more rows
```

A first glance at the table shows that there are clear differences between the samples (within a liking variable), but little difference between liking variables (within a sample).

Of course, we want to know if differences between samples are significant. We thus need to perform a two-way ANOVA (testing for the product effect and also taking into account the individual differences) followed up by a paired comparison test (here Tukey's HSD). For the latter, the {`agricolae`} package is a good solution, as it is simple to use and has all its built-in tests working in the same way.

```
library(agricolae)

liking_start <- lm(`1stbite_liking` ~ Product + Judge, data = consumer)
liking_start_hsd <- HSD.test(liking_start, "Product")$groups %>%
  as_tibble(rownames = "Product")

liking_start_hsd
```

```
## # A tibble: 10 x 3
##   Product '1stbite_liking' groups
##   <chr>                <dbl> <chr>
## 1 P10                   7.40 a
## 2 P1                    6.30 b
## 3 P5                    5.78 b
## 4 P2                    5.53 bc
## # ... with 6 more rows
```

```
liking_end <- lm(`after_liking` ~ Product + Judge, data = consumer)
liking_end_hsd <- HSD.test(liking_end, "Product")$groups %>%
  as_tibble(rownames = "Product")
```

At both evaluation times, the Tukey's HSD test shows significant differences in liking between samples (at a 5% risk).

To further compare the liking assessment of the samples after the first bite and at the end of the tasting, the results obtained from `liking_start_hsd` and `liking_end_hsd` are combined. We then represent the results in a bar chart (Figure 10.9):

```
list(Start = liking_start_hsd %>% rename(Liking = `1stbite_liking`),
     End = liking_end_hsd %>% rename(Liking = `after_liking`)) %>%
  enframe(name = "Moment", value = "res") %>%
  unnest(res) %>%
  mutate(Moment = factor(Moment, levels = c("Start", "End"))) %>%
  ggplot(aes(x = reorder(Product, -Liking), y = Liking, fill = Moment))+
  geom_bar(stat = "identity", position = "dodge")+
  xlab("")+
  theme_bw()
```

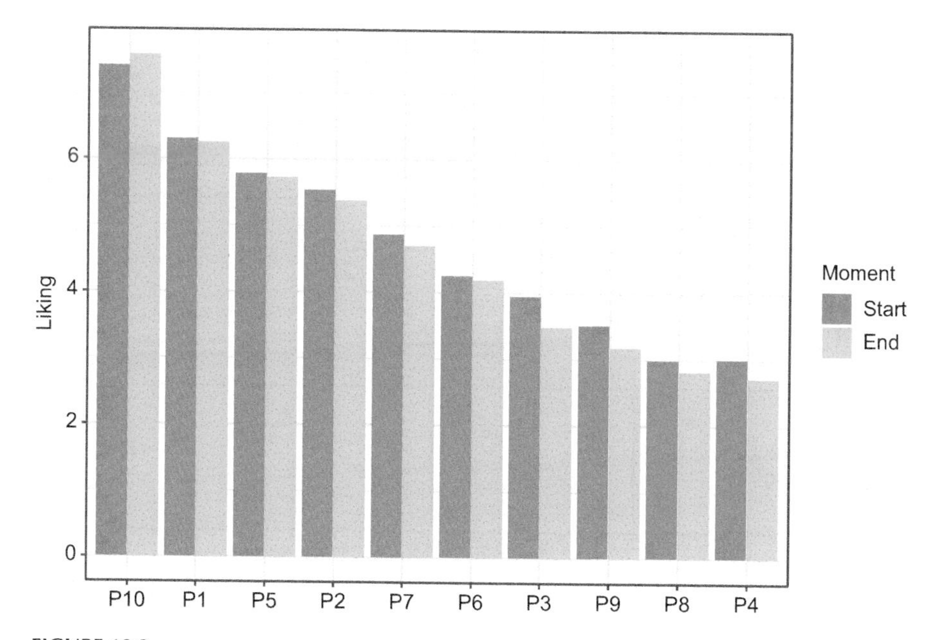


**FIGURE 10.9**
Bar chart for the the comparison of mean liking scores at the first bite and at the end of the evaluation.

As can be seen, the pattern of liking scores across samples is indeed very stable across the evaluation, particularly in terms of rank. At the individual level, such linear relationship is also observed (here for the first 12 consumers) (Figure 10.10):

```
consumer %>%
  dplyr::select(Judge, Product, Start = `1stbite_liking`,
                End = `after_liking`) %>%
  filter(Judge %in% str_c("J", 1:12)) %>%
  mutate(Judge = factor(Judge, levels = unique(
    str_sort(.$Judge, numeric = TRUE)))) %>%
  ggplot(aes(x = Start, y = End)) +
  geom_point(pch = 20, cex = 2) +
  geom_smooth(method = "lm", formula = "y~x", se = FALSE) +
  theme_bw() +
  facet_wrap(~Judge)
```

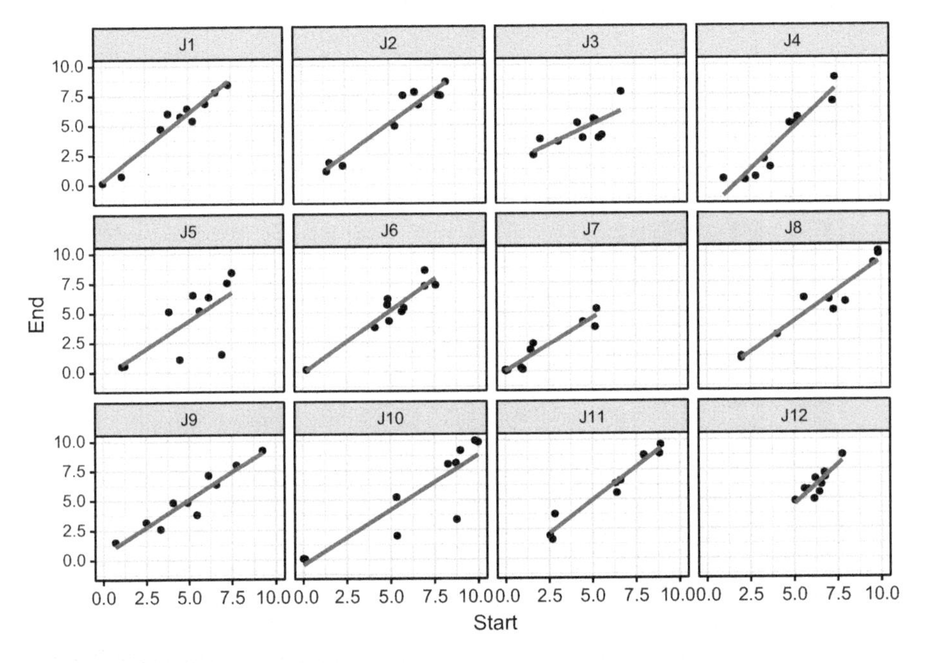


**FIGURE 10.10**
Faceted scatterplots showing the linear relationship between overal liking at the first bite vs. end of the tasting for the first 12 consumers.

---

For your own curiosity, we invite you to recreate the same graph by comparing the Liking score at the end of the evaluation (`after_liking`) with the liking score measured on the 9pt categorical scale (`end_liking 9pt`) and to reflect on the results obtained. Are the consumers consistent in their evaluations?

---

Another interesting relationship to study involves the liking scores[4] and the number of cookies eaten by each consumer. We could follow the same procedure as before, but prefer to add here a filter to only show consumers with a significant regression line at 5%.

Let's start by creating a function called `run_reg()` that runs the regression analysis of the number of biscuits (`N`) in function of the liking score (`Liking`):

```
run_reg <- function(df) {
  output <- lm(N ~ Liking, data = df)
  return(output)
}
```

After transforming the data, we apply this function to each consumer separately.

Here, we take advantage of the flexibility of tibbles as it allows storing results as list by saving three sorts of outputs per consumer:

- `data` contains the individual data;
- `lm_obj` corresponds to the results of the linear model (obtained with 'run_reg()"');
- `glance` contains some general results of the model incl. $R^2$, p-value, etc.

---

These three output contain completely different information for the same analysis (here regressions). The fact that tibbles allow storing output as list is very handy since all the results are tidied in one unique R object, which can then easily be accessed by unfolding the output needed.

---

```
N_liking_reg <- consumer %>%
  dplyr::select(Judge, Product, Liking = `end_liking 9pt`, N) %>%
  mutate(Liking = 10 - Liking) %>%
  group_by(Judge) %>%
  nest() %>%
  ungroup() %>%
  mutate(lm_obj = map(data, run_reg)) %>%
  mutate(glance = map(lm_obj, broom::glance))
```

[4] We would like to remind the reader that the liking scores measured on the categorical scale was reverted since 1 defined "I like it a lot" and 9 "I dislike it a lot". To simplify the readability, this scale is reverted so that 1 corresponds to a low liking score and 9 to a high liking score (in practice, we will take as value 10 – score given).

Since we only want to represent consumers with a significant regression line, we unfold the results stored in `glance` so that we can access the `p.value` of each regression.

```
N_liking <- N_liking_reg %>%
  unnest(glance) %>%
  filter(p.value <= 0.05) %>%
  arrange(p.value) %>%
  mutate(Judge = fct_reorder(Judge, p.value)) %>%
  unnest(data)
```

Ultimately, the relationship between the liking score and the number of biscuits eaten is represented in a line chart (Figure 10.11):

```
ggplot(N_liking, aes(x = Liking, y = N)) +
  geom_point(pch = 20, cex = 2) +
  geom_smooth(method = "lm", formula = "y~x", se = FALSE) +
  theme_bw() +
  scale_y_continuous(labels = label_number(accuracy = 1)) +
  ggtitle("Number of Biscuits vs. Liking",
          "Consumers with a signif. (5%) regression model are shown.") +
  facet_wrap(~Judge)
```

## 10.4 Combining Sensory and Consumer Data

### 10.4.1 Internal Preference Mapping

Now that we have analyzed the sensory and the consumer data separately, it is time to combine both data sets and analyze them conjointly. A first analysis that can then be performed is the internal preference mapping, that is, a PCA on the consumer liking scores in which the sensory attributes are projected as supplementary (Figures 10.12 and Figure 10.13).

Such analysis is split into three steps:

1. The consumer data are reorganized in a wide format with the samples in rows and the consumers in columns;
2. The sensory mean table is joined to the consumer data (make sure that the product names perfectly match in the two files);
3. A PCA is performed on the consumer data, with the sensory descriptors being projected as supplementary variables.

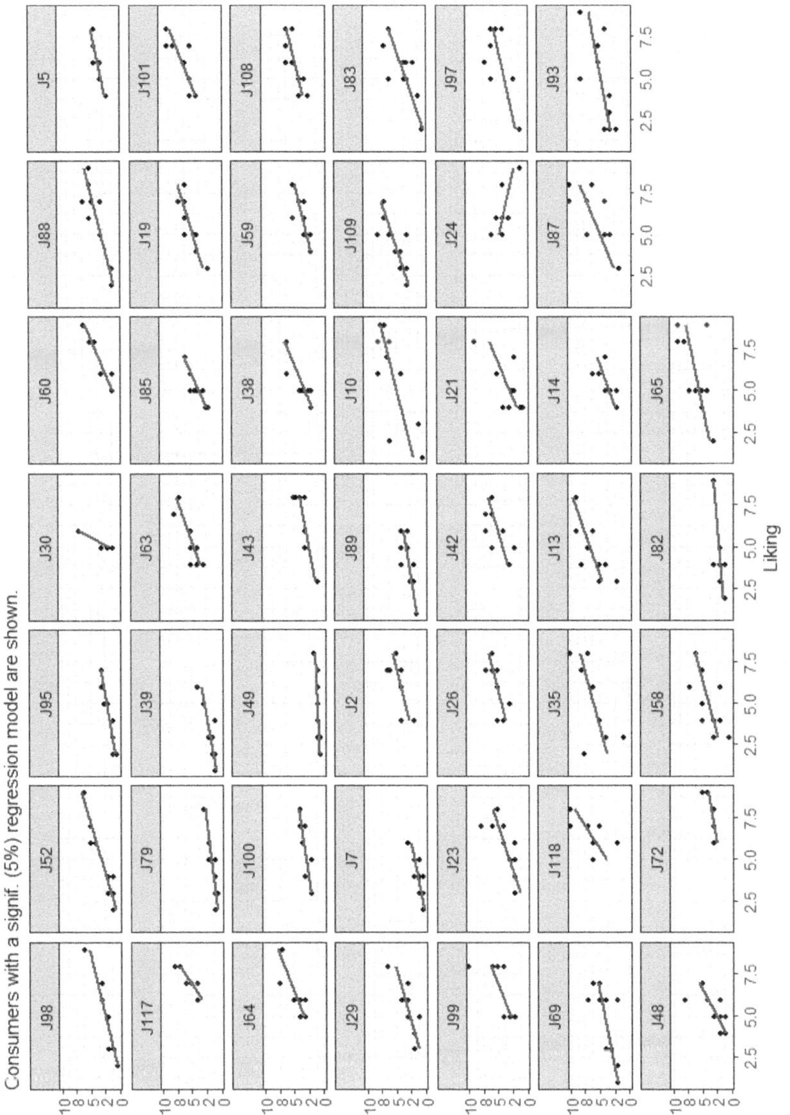


**FIGURE 10.11**
Consumers showing a significant relationship between their liking scores and the number of biscuits they ate.

```
consumer_wide <- consumer %>%
  separate(Product, into = c("P", "Number"), sep = 1) %>%
  mutate(Number = ifelse(nchar(Number) == 1,
                         str_c("0", Number), Number)) %>%
  unite(Product, P, Number, sep = "") %>%
  dplyr::select(Judge, Product, Liking = `end_liking 9pt`) %>%
  mutate(Liking = 10 - Liking) %>%
  pivot_wider(names_from = Judge, values_from = Liking)

data_mdpref <- senso_mean %>%
  inner_join(consumer_wide, by = "Product")

res_mdpref <- data_mdpref %>%
  as.data.frame() %>%
  column_to_rownames(var = "Product") %>%
  PCA(., quali.sup = 1:2, quanti.sup = 3:34, graph = FALSE)

fviz_pca_ind(res_mdpref, habillage = 1)
```

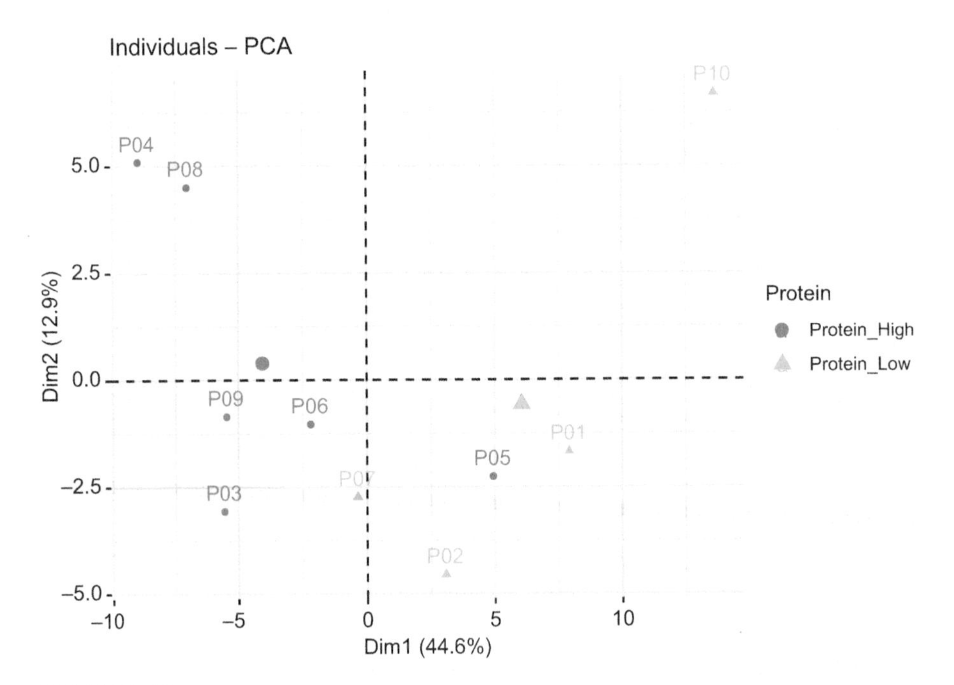


**FIGURE 10.12**
Internal preference mapping (PCA score plot of consumer data).

```
fviz_pca_var(res_mdpref, label = "quanti.sup",
             select.var = list(cos2 = 0.5), repel = TRUE)
```

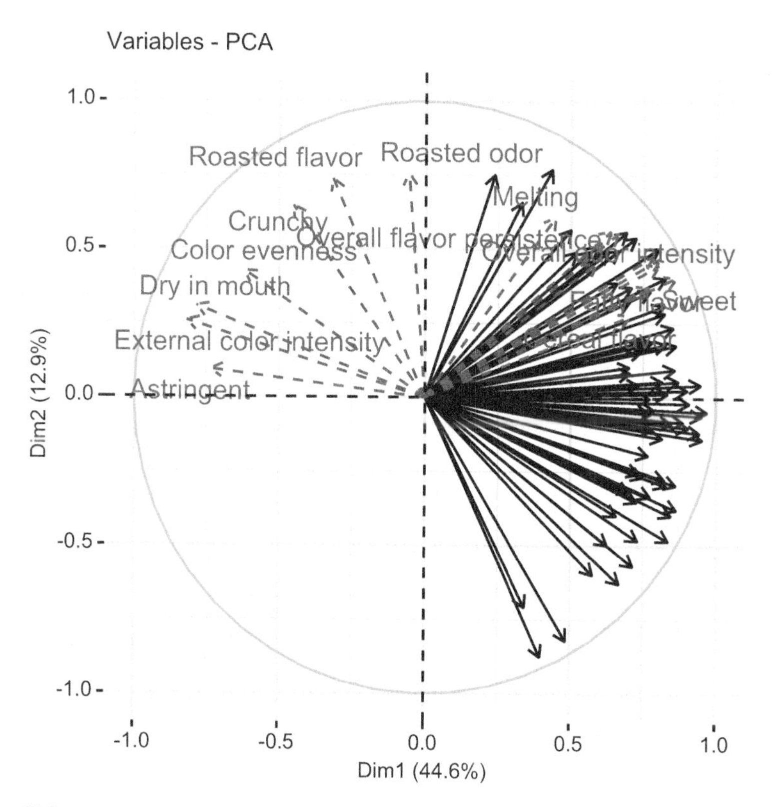


**FIGURE 10.13**
Internal preference mapping (PCA loadings of consumer data with sensory descriptors as supplementary variables).

As can be seen, the consumers are quite in agreement as all the black arrows are pointing in a similar direction. Overall, they seem to like biscuits that are sweet, with cereal flavor, and fatty/dairy flavor and odor, and dislike biscuits defined as astringent, dry in mouth, uneven, and with dark external color.

### 10.4.2 Consumers Clustering

Even though the data show a fairly good agreement between consumers, let's cluster them in more homogeneous groups based on liking.

There are various solutions for clustering depending on the type of distance (similarity or dissimilarity), the linkage (single, average, Ward, etc.), and of course the algorithm itself (e.g. AHC, k-means, etc.).

Here, we opt for Agglomerative Hierarchical Clustering (AHC) with Euclidean distance (dissimilarity) and Ward criterion, as it is a fairly common approach in Sensory and Consumer research. Such analysis can be done using `stats::hclust()` or `cluster::agnes()`.

---

Before computing the distance between consumers, it is advised to center their liking scores (subtracting their mean liking scores to each of their individual scores) in order to group consumers based on their respective preferences, rather than on their scale usage (otherwise, consumers who scored high on all samples are grouped together and separated from consumers who scored low on all samples, which isn't so much informative). Some researchers recommend to also scale the liking scores. Both transformations can be done automatically using the `scale()`[5] function.

---

Let's start with computing the Euclidean distance between each pair of consumers by using the `dist()` function.

```
consumer_dist <- consumer_wide %>%
  as.data.frame() %>%
  column_to_rownames(var = "Product") %>%
  scale(., center = TRUE, scale = FALSE) %>%
  t(.) %>%
  dist(., method = "euclidean")
```

The AHC is performed using the `hclust()` function and the `method = "ward.D2"` parameter, which is the equivalent to `method = "ward"` for `agnes()`. To visualize the resulting dendrogram, we can use the `factoextra::fviz_dend()` function (here we propose to visualize the two-clusters solution by setting `k=2`) (Figure 10.14):

```
res_hclust <- hclust(consumer_dist, method = "ward.D2")
fviz_dend(res_hclust, k = 2)
```

[5] `scale()` allows centering (`center=TRUE`) and standardizing (`scale=TRUE`) data automatically in columns, hence generating *z-scores*.

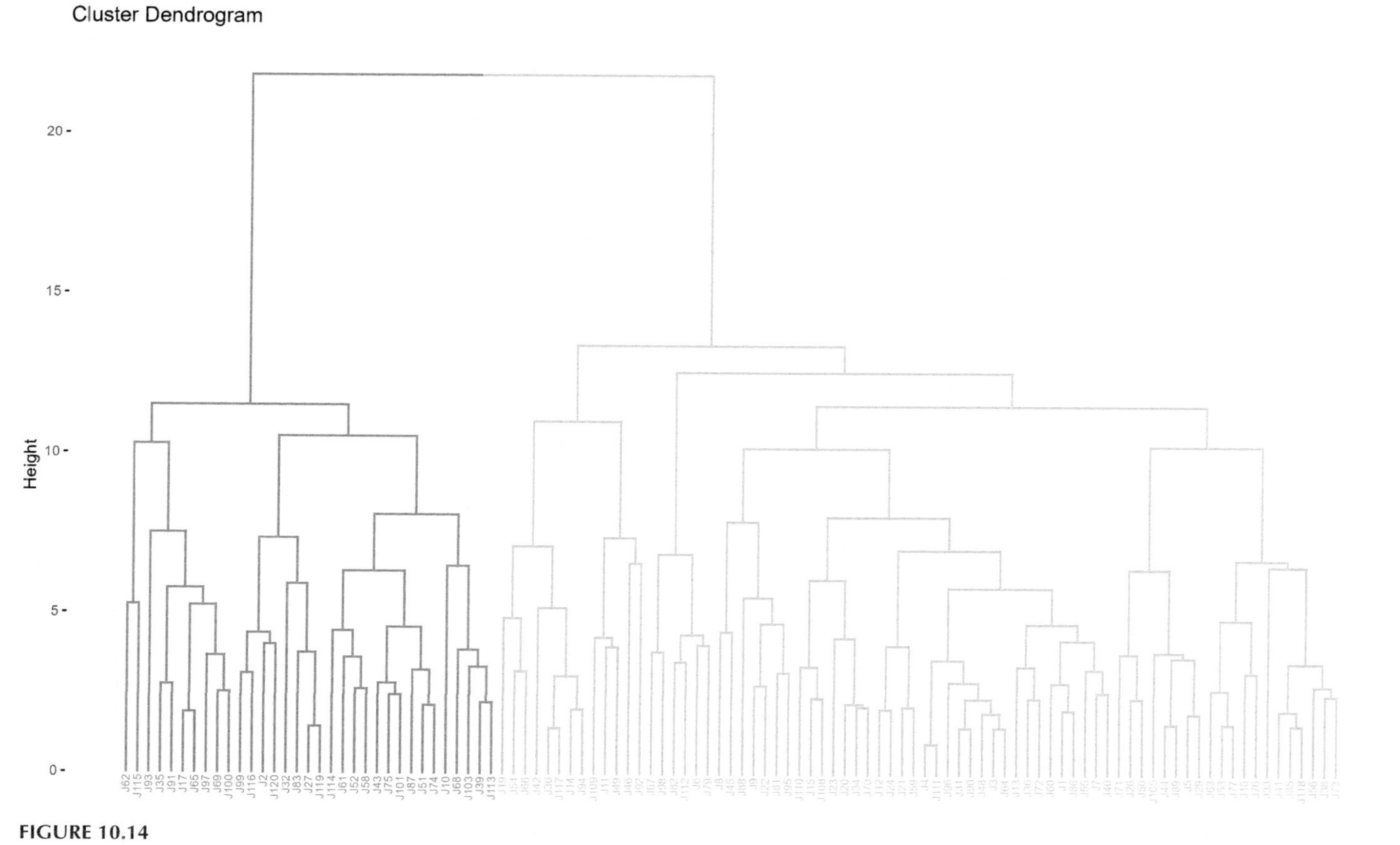


**FIGURE 10.14**
Agglomerative hierarchical clustering of consumers' centered liking scores.

---

An interesting option to visualize clusters and proposed by `fviz_dend()` is the *phylogenic* representation (`type="phylogenic"`). We invite you to give it a try to see how it represents the clusters as an alternative to the classical dendrogram tree.

---

The two clusters solution looks satisfactory, so we can cut the tree at this level (using `cutree()`), hence generating a group of 74 and a group of 33 consumers.

```
res_clust <- cutree(res_hclust, k = 2) %>%
  as_tibble(rownames = "Judge") %>%
  rename(Cluster = value) %>%
  mutate(Cluster = as.character(Cluster))

res_clust %>%
  count(Cluster)
```

```
## # A tibble: 2 x 2
##   Cluster     n
##   <chr>   <int>
## 1 1          74
## 2 2          33
```

Lastly, we can compare visually the preference patterns of the two clusters by representing in a line chart the average liking score for each product provided by each cluster (Figure 10.15).

```
mean_cluster <- consumer %>%
  separate(Product, into = c("P", "Number"), sep = 1) %>%
  mutate(Number = ifelse(nchar(Number) == 1,
                         str_c("0", Number), Number)) %>%
  unite(Product, P, Number, sep = "") %>%
  dplyr::select(Judge, Product, Liking = `end_liking 9pt`) %>%
  mutate(Liking = 10 - Liking) %>%
  full_join(res_clust, by = "Judge") %>%
  group_by(Product, Cluster) %>%
  summarize(Liking = mean(Liking), N = n()) %>%
  mutate(Cluster = str_c(Cluster, " (", N, ")")) %>%
  ungroup()

ggplot(mean_cluster, aes(x = Product, y = Liking,
                         colour = Cluster, group = Cluster)) +
  geom_point(pch = 20) +
  geom_line(aes(group = Cluster), lwd = 2) +
```

```
xlab("") +
scale_y_continuous(name = "Average Liking Score",
                   limits = c(1, 9), breaks = seq(1, 9, 1)) +
theme_bw()
```

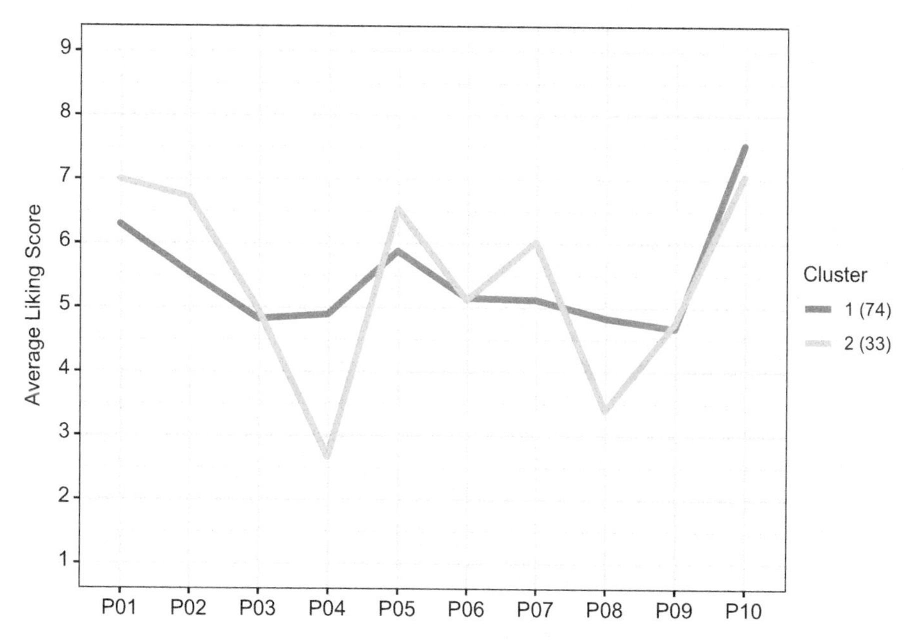


**FIGURE 10.15**
Average liking scores of the two consumer clusters.

It appears that cluster 1 (74 consumers) particularly likes `P10`, `P01`, and `P05`, and has a fairly flat liking pattern otherwise. On the other hand, cluster 2 (33 consumers) expressed strong rejections toward `P04` and `P08`, and liked `P10` and `P01` the most.

The fact that both clusters agree on the best samples (`P10` and `P01`) goes with our original observation from the Internal Preference Mapping that the panel of consumers is fairly homogeneous in terms of preferences.

---

In {`FactoMineR`}, the `HCPC()` function also performs AHC but takes the results of a multivariate analysis as a starting point (HCPC stands for Hierarchical Clustering on Principal Components). This would typically be the results of the PCA performed on the $Consumer(rows) \times Product(columns)$ matrix of liking scores, in which the scores are (at least) centered in row. Although results should be identical in most cases, it can happen that results slightly diverge from `agnes()` and `hclust()` as it also depends on the number of dimensions kept in the multivariate analysis and on the treatment of in-between clusters consumers. But more interestingly, `HCPC()` offers the possibility to *consolidate* the clusters by performing k-means on the solution obtained from the AHC (`consol=TRUE`).

---

### 10.4.3 Drivers of Liking

Combining sensory and consumer data collected on the same products allows understanding which sensory properties of the products drive the consumers' liking and disliking. Such evaluation can be done at the panel level, at a subgroup level (e.g. clusters, users vs. non-users, gender, etc.), or even at the individual level. Unless stated otherwise, the computations will be done for cluster 1, but could easily be adapted to other groups if needed.

#### *Correlation*

Let's start by evaluating the simplest relationship between the sensory attributes and overall liking by looking at correlations. Here, we are combining the average liking score per cluster to the sensory profile of the products. The correlations are then computed using the `cor()` function:

```
data_cor <- mean_cluster %>%
  dplyr::select(-N) %>%
  pivot_wider(names_from = Cluster, values_from = Liking) %>%
  inner_join(senso_mean %>%
               dplyr::select(-c(Protein,Fiber)), by="Product") %>%
  as.data.frame() %>%
  column_to_rownames(var = "Product")

res_cor <- cor(data_cor)
```

Various packages can be used to visualize these correlations. We opt here for `ggcorrplot()` from the {ggcorrplot} package as it provides many interesting visualization options based on {ggplot2} (Figure 10.16). This package also provides the function `cor_pmat()` which computes the p-value associated with each correlation. This matrix of p-value can be used to hide correlations that are not significant at the level defined by the parameter `sig.level`.

```
library(ggcorrplot)
res_cor_pmat <- cor_pmat(data_cor)
```

```
ggcorrplot(res_cor, type = "full", p.mat = res_cor_pmat,
           sig.level = 0.05, insig = "blank",
           lab = TRUE, lab_size = 1.5)
```

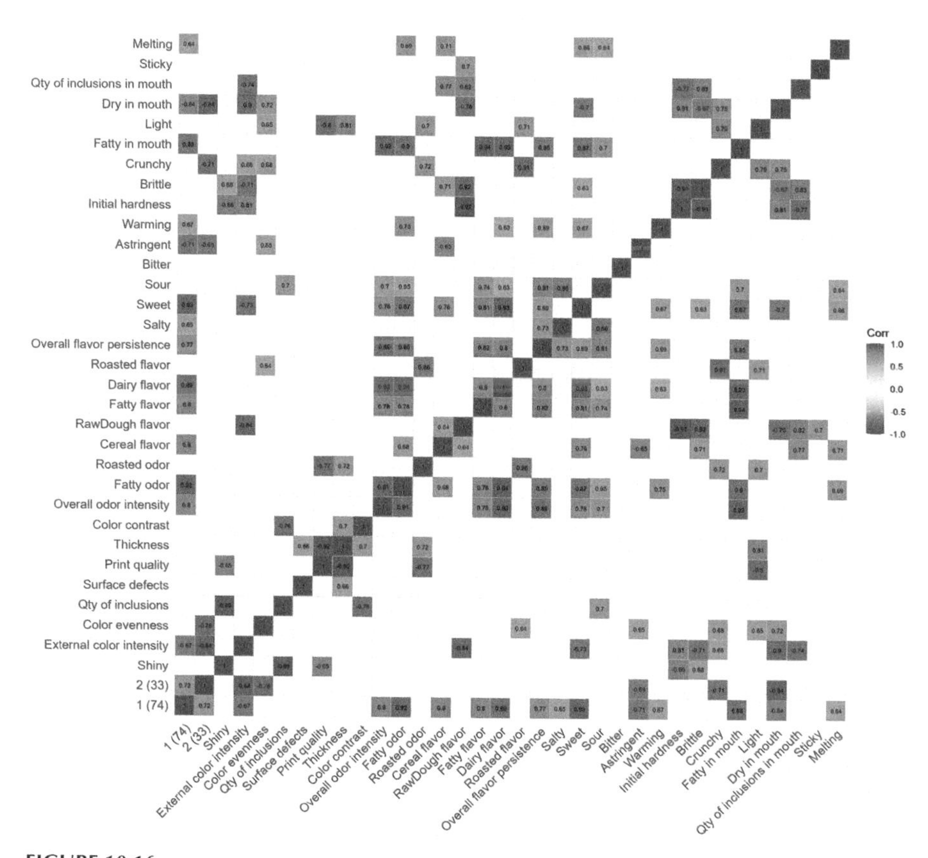


**FIGURE 10.16**
Matrix of pairwise correlations between attributes.

The average liking scores for cluster 1 (defined as `1 (74)`) are positively correlated with `Overall odor intensity`, `Fatty odor`, `Cereal flavor`, `Fatty flavor`, `Dairy flavor`, `Overall flavor persistence`, `Salty`, `Sweet`, `Warming`, `Fatty in mouth`, and `Melting`. They are also negatively correlated to `External color intensity`, `Astringent`, and `Dry in mouth`. Finally, it can be noted that the correlation between clusters is high with a value of 0.72.

### Linear and Quadratic Regression

Although the correlation provides a first good idea of which attributes are linked to liking, it only measures linear relationships and it does not allow for inference. To overcome these limitations, linear and quadratic regressions are used.

Let's start by combining the sensory data to the average liking score per product for cluster 1. To simplify the analysis, all the sensory attributes are structured in the longer format.

```
data_reg <- mean_cluster %>%
  filter(Cluster == "1 (74)") %>%
  dplyr::select(-N) %>%
  inner_join(senso_mean %>%
             dplyr::select(-c(Protein, Fiber)), by = "Product") %>%
  pivot_longer(Shiny:Melting,
               names_to = "Attribute", values_to = "Score") %>%
  mutate(Attribute =
           factor(Attribute,
                  levels=colnames(senso_mean)[4:ncol(senso_mean)]))
```

Both the simple linear regression and the quadratic regression are then run on `Liking` per attribute:

---

To add a quadratic model, two options are possible: 1. In `data_reg`, we could add a column (using `mutate()`) called `Score2` that is defined as `Score2 = Score^2`. The model for the quadratic regression is then defined as `Liking ~ Score + Score2`; 2. The quadratic model is informed directly using the `poly()` function by informing which polynomial degrees to consider (here `2`). For its concision, we opt for the second option.

---

```
res_reg <- data_reg %>%
  nest_by(Attribute) %>%
  mutate(lin_mod = list(lm(Liking ~ Score, data=data)),
         quad_mod = list(lm(Liking ~ poly(Score, 2), data=data)))
```

We extract the attributes that are significantly linked to liking (at 5%, and we accept 6% for quadratic effects). To do so, the results stored in `lin_mod` and `quad_mod` need unfolding (`summarize()`) and restructuring (`broom::tidy()`).

```
lin <- res_reg %>%
  reframe(broom::tidy(lin_mod)) %>%
  ungroup() %>%
  filter(term == "Score", p.value <= 0.05) %>%
  pull(Attribute) %>%
  as.character()

quad <- res_reg %>%
  reframe(broom::tidy(quad_mod)) %>%
  ungroup() %>%
  filter(term == "poly(Score, 2)2", p.value <= 0.06) %>%
  pull(Attribute) %>%
  as.character()
```

These attributes are then represented graphically against the liking Scores (Figure 10.17).

```
library(ggrepel)

df <- data_reg %>%
  filter(Attribute %in% unique(c(lin, quad)))

p <- ggplot(df, aes(x = Score, y = Liking, label = Product)) +
  geom_point(pch = 20, cex = 2) +
  geom_text_repel() +
  theme_bw() +
  facet_wrap(~Attribute, scales = "free_x")
```

Let's now add a regression line to the model. To do so, `geom_smooth()` is being used with as `method = lm` combined to `formula = 'y ~ x'` for linear relationships, and `formula = 'y ~ x + I(x^2)'` for quadratic relationships (when both the linear and quadratic models are significant, the quadratic model is used).

```
lm.mod <- function(df, quad) {
  ifelse(df$Attribute %in% quad, "y~x+I(x^2)", "y~x")
}
```

We apply this function to our data by applying to each attribute (here we set `se=FALSE` to remove the confidence intervals around the regression line):

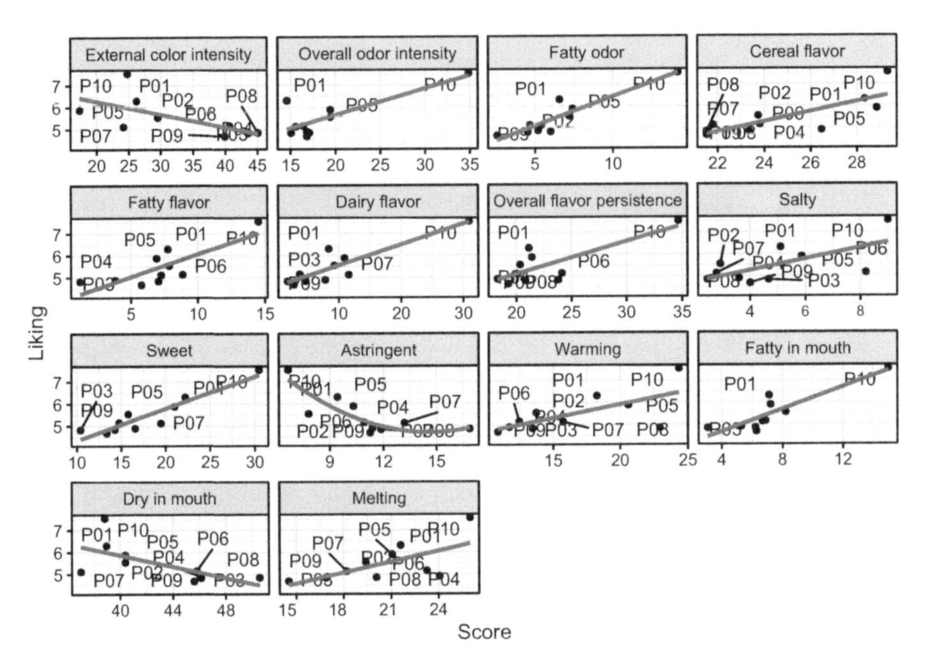


**FIGURE 10.17**
Regression of liking on sensory atttributes taken separately.

```
p_smooth <- by(
  df, df$Attribute,
  function(x) geom_smooth(data = x, method = lm,
                          formula = lm.mod(x, quad = quad),
                          se = FALSE)
)
p + p_smooth
```

All attributes except `Astringent` are linearly linked to liking. For `Astringent`, the curvature is U-shaped: This does not show an effect of saturation as it would have been represented as an inverted U-shape. Although the quadratic effect shows a better fit than the linear effect, having a linear effect would have been a good predictor as well in this situation.

Such an approach to regression is quite straightforward, however it must be applied and interpreted with caution because of potential multicollinearity among predictor variables. One way around this issue is to take all the sensory attributes together and perform a regression on principal components as described in Section 10.4.4. Another option would be to run a partial least square regression (see also Section 12.5).

### 10.4.4 External Preference Mapping

Ultimately, one of the goals of combining sensory and consumer data is to find areas within the sensory space that are best liked by consumers. Since this approach is based on modeling and prediction, it may suggest areas of the sensory space with high acceptance potential which are not yet filled in by existing products. This would thus open doors to new product development.

To perform such an analysis, one option is to use External Preference Mapping (PrefMap) (Schlich and McEwan, 1992; Danzart, 1998; and Danzart et al., 2004).[6]

To run the PrefMap analysis, the `carto()` function from {`SensoMineR`} is being used. This function takes as parameters the sensory space to consider (stored in `senso_pca$ind$coord`, here we will consider dimension 1 and dimension 2), the table of liking scores (as stored in `consumer_wider`), and the model to consider (here we consider the quadratic model, so we use `regmod=1`). For convenience, we run the analysis on the full panel since `consumer_wider` is (almost) already structured as needed.

Note that `carto()` requires a matrix or data frame with row names for the analysis. Thus, the data need to be slightly adapted (we also need to ensure that the products are in the same order in both files).

```
senso <- senso_pca$ind$coord[, 1:2] %>%
  as_tibble(rownames = "Product") %>%
  arrange(Product) %>%
  as.data.frame() %>%
  column_to_rownames(var = "Product")

consu <- consumer_wide %>%
  arrange(Product) %>%
  as.data.frame() %>%
  column_to_rownames(var = "Product")

nconso = ncol(consu)

library(SensoMineR)
PrefMap <- carto(Mat = senso, MatH = consu, regmod = 1,
                 graph.tree = FALSE, graph.corr = FALSE,
                 graph.carto = TRUE)
```

From this map, we can see that the optimal area (dark red) is located on the positive side of dimension 1, between `P01`, `P05`, and `P10` (as expected by the liking score).

[6] For more information on the principles of PrefMap, please refer to MacFie (2007), Meullenet et al. (2008), and Lê and Worch (2018).

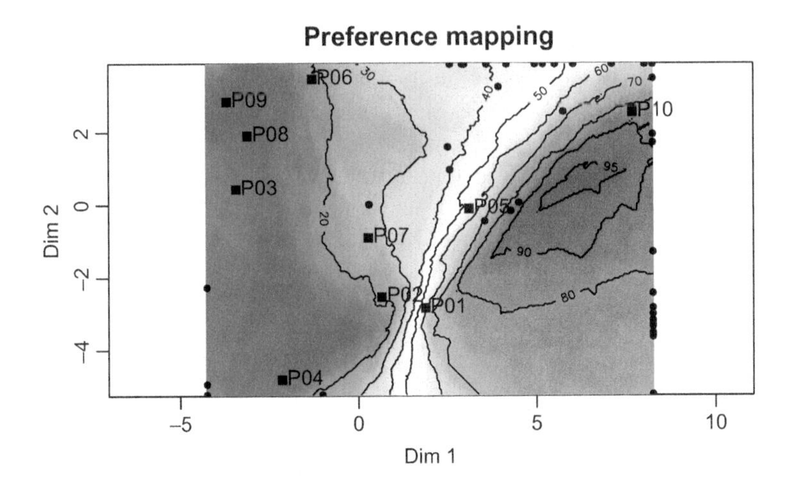


Let's now rebuild this plot using {ggplot2}.

The sensory space is stored in senso, whereas the response surface plot information is split between:

- PrefMap$f1: contains the coordinates on dimension 1 in which predictions have been made;
- PrefMap$f2: contains the coordinates on dimension 2 in which predictions have been made;
- PrefMap$nb.depasse: contains the number of consumers at each point of the space that would accept a product located at that point. This matrix is defined in such a way that PrefMap$f1 links to the rows of the matrix, and PrefMap$f2 links to the columns. Note that this counts should be transformed into percentages by dividing each count by the number of consumers who are involved in the test (defined earlier as nconso)

Last but not least, POpt (whose coordinates are stored in senso_pca$ind.sup$coord) can be projected on that map in order to see how this optimized sample is considered in terms of consumers' preference.

Let's start with preparing the data by transforming everything back into a tibble:

```
senso <- senso %>%
  as_tibble(rownames = "Product")
senso_sup <- senso_pca$ind.sup$coord %>%
  as_tibble(rownames = "Product")
```

```
dimnames(PrefMap$nb.depasse) <- list(round(PrefMap$f1, 2),
                                     round(PrefMap$f2, 2))
PrefMap_plot <- PrefMap$nb.depasse %>%
  as_tibble(rownames = "Dim1") %>%
  pivot_longer(-Dim1,
               names_to = "Dim2", values_to = "Count") %>%
  mutate("Preference (%)" = Count * 100 / nconso) %>%
  mutate(across(where(is.character), as.numeric))
```

To build the plot, different layers involving different source of data (`senso`, `senso_sup`, and `PrefMap_plot` that is) are required. Hence, the initiation of the plot through `ggplot()` does not specify any data. Instead, the data used in each step are included within the `geom_*()` of interest. In this example, `geom_tile()` (coloring) and `geom_contour()` (contour lines) are used to build the surface plot (Figure 10.18).

```
ggplot() +
  geom_tile(data = PrefMap_plot, aes(x = Dim1, y = Dim2,
                                     fill = `Preference (%)`,
                                     color = `Preference (%)`)) +
  geom_contour(data = PrefMap_plot, aes(x = Dim1, y = Dim2,
                                        z = `Preference (%)`),
               breaks = seq(0, 100, 10)) +
  geom_hline(yintercept = 0, lty = 2) +
  geom_vline(xintercept = 0, lty = 2) +
  geom_point(data = senso, aes(x = Dim.1, y = Dim.2),
             pch = 20, cex = 3) +
  geom_text_repel(data = senso, aes(x = Dim.1, y = Dim.2,
                                    label = Product)) +
  geom_point(data = senso_sup, aes(x = Dim.1, y = Dim.2),
             pch = 20, col = "white", cex = 3) +
  geom_text_repel(data = senso_sup, aes(x = Dim.1, y = Dim.2,
                                        label = Product),
                  col = "white") +
  scale_fill_gradient2(low = "blue", mid = "white", high = "red",
                       midpoint = 50) +
  scale_color_gradient2(low = "blue", mid = "white", high = "red",
                        midpoint = 50) +
  xlab(str_c("Dimension 1(", round(senso_pca$eig[1, 2], 1), "%)")) +
  ylab(str_c("Dimension 2(", round(senso_pca$eig[2, 2], 1), "%)")) +
  ggtitle("External Preference Mapping applied to the biscuits data",
          "(The PrefMap is based on the quadratic model)") +
  theme_bw()
```

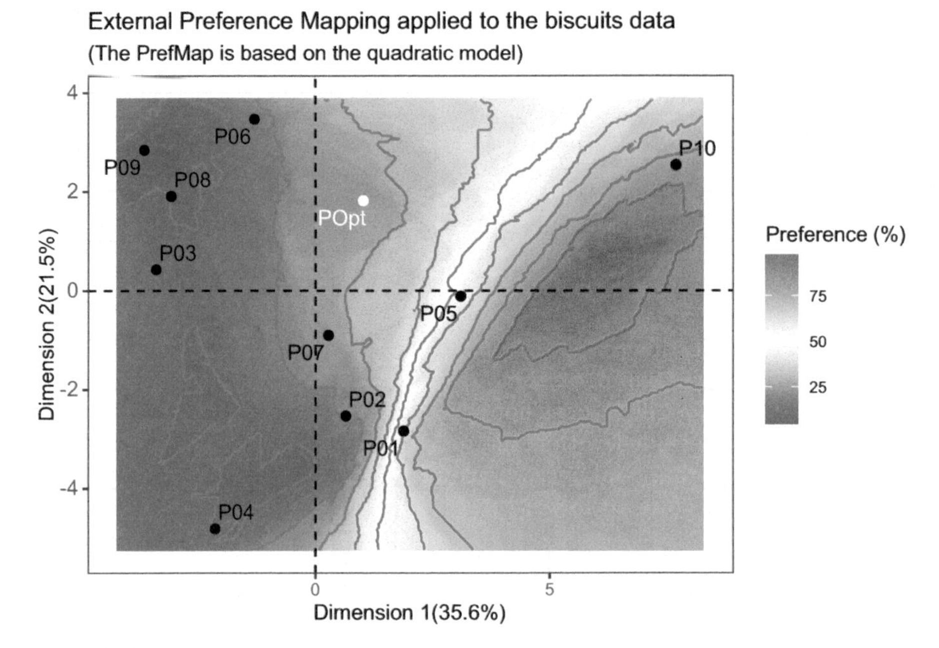


**FIGURE 10.18**
Plot of the External Preference Mapping using ggplot2.

As can be seen, `POpt` is quite far from the optimal sensory area suggested by the PrefMap. This suggests that prototypes with higher success chances could be developed.

# 11

# *Value Delivery*

Most of this book focus on handling data through the procedure of data cleaning, transformation, analyses, or representation. This makes sense since the aim is to present data science in the context of sensory and consumer research. But these steps may be irrelevant if analysts or reseachers are not able to communicate their findings efficiently. Effective communication is as much important as any other step and so, we will guide you through that, providing some valuable tips and practical examples. This chapter approaches important topics to help you reach the goal of successful communication, from understanding the distinction between different analysis and audience awareness to methods to communicate, the art of storytelling, and ultimately, reformulate.

## 11.1 How to Communicate?

Sensory and consumer scientists often act as consultant whether it be for their own company or for customers. Being able to communicate effectively is perhaps one of the most important skills they should master. Communication is a simple act of transferring information and although undervalued by many, plays a key role in any business's success. Let's start this chapter reminding that there are different ways to communicate, and this process usually includes a combination of two or more types of languages, which are:

- **Vocal** – the language produced by articulate sounds. It is the language used in clients' meetings and presentations for instance.
- **Non-verbally** – is related to the body language, gestures, and the tone and pitch of the voice.
- **Written** – is the representation of a spoken language in a writing system, like proposals, technical documents, or final reports.

DOI: 10.1201/9781003028611-11

- **Visual** – is the communication using visual elements, such as the visual quality of presentations or other written documents, including formatting, logo, colors, figures, plots, etc.

It is through effective communication that you will bring potential clients' attention and interest in your company and the services you provide, that will make you truly understand your clients' needs, gain their trust, and provide the right solutions that will ultimately bring to a long-term partnership. Efficacious communication will be responsible for keeping a friendly relationship and your clients' commitment throughout the project development and will also help you to properly convey the outcomes of a project in a way that will at least meet (or better surpass) your clients' expectations and opens possibilities for follow-up engagements and/or recommendations.

The ability to communicate accurately, clearly, and as intended is definitely something that consultants should not overlook because although it seems straightforward, it involves a number of skills that may take several years of practice to master. You will find plenty of materials on the Internet and books to help you to understand better and develop your very basic skills for effective communication. We will not focus on that in this chapter, but it is worth highlighting some important aptitudes for vocal communication, which may configure one of the most powerful types of communication with your client:

- **Confidence**: Being confident makes you to be seen as an expert on the topic and as having the situation under control. The audience will be more likely to trust, believe, be connected, and give credit to a confident person.
- **Passion and Enthusiasm**: Be passionate about what you do and convinced/enthusiastic about the solution you provide. The audience can easily capture that on your vocal and non-verbal language and will be much more interested if they can see and feel your passion.
- **Ability to be succinct**. No matter how interesting you feel about a topic, you must know that the audience will lose interest after some time, especially if there is a lot of technical and detailed information. Be aware that the attention span of your audience is not long, so use your time wisely during the presentation, keeping it short and at the point.
- **Feeling**. This is a skill that definitely one needs time to master, but it is crucial that you pick what is going on with your client, if they seem to be understanding and following you or if they seem to be confused or not sure about what you are talking about.

In this chapter, we focus on four topics that we believe any successful consultant should have in mind which are: Exploratory, Explanatory, and Predictive Analysis; Audience Awareness; Method to Communicate; and Storytelling.

## 11.2 Exploratory, Explanatory, and Predictive Analysis

As a consultant in the field, you are likely in a position where you get data from your client and review it, do the analyses, and ultimately, convey the results. And here is where it is important to make a clear distinction between exploratory and explanatory analysis. **Exploratory analysis** is the stage where you dig into the data, get to understand them, figure out patterns and things that may be interesting or important to highlight. **Explanatory analysis** is the ability, from the learning from the previous step, to select and/or reorganize your data (by remaking your tables, plots, or charts) in a way you can easily convey the message to your audience, and ultimately make them understand and focus on the things that are worthy. Let's discuss that a little bit more.

The hard work starts once you get the data. This is the time you will likely analyze it in multiple ways, make several plots, and look at the data from multiple angles. This is what we call, exploratory analysis! After understanding all the analysis, it may be tempting to show the audience everything, all the steps, decisions, different plots, and approaches you have taken, but **do not do that**. You do not want to overload your audience making them go through the same tough path you went. Instead of showing your handwork, the robustness of your analysis, and building up your credibility, you have to make your audience confused, bored, and lacking interest.

Once you have done all the hard work on data analysis, it is the moment to take some time to stand back and look at the key findings and the message(s) you want to convey. It is important to keep in mind that there is always a balance to find between presenting quantified, accurate, and credible information (i.e. with sufficient details) and presenting information that makes sense, is relevant, and that is easily readable and understandable. This challenging phase is what we call explanatory analysis! This is the moment when you need to use your ability to translate an extensive, detailed, and complex version of your data analysis to a more concise/holistic version, to a version that will easily and clearly convey the message, and highlight the main points. Keep in mind that the explanatory analysis has to be tailored according to your audience (as discussed in Section 11.3), which means that the way you present the data analysis and the level of details provided vary if you are presenting it to a group of experts in the field, including statisticians and mathematicians or in a lecture for a very diverse audience, as in a conference for instance. You need to find the right balance!

Some examples in the field to exemplify the two extremes (too complicated or too simple):

- *Factorial maps* – The overuse of factorial maps is a common practice in the sensory and consumer science field. It is a great tool to explore data, to make or confirm hypotheses, but may not be the best to communicate since not so many people can correctly read and interpret them. Therefore, a good approach would be to initially work with the factorial map to interpret and draw conclusions, but then, find another way through tables or alternative charts (that may be simpler to understand) to communicate the findings to your audience.

- *Spider Plots* – This is the other extreme when consultants can fail but not because they present a very complex and extensive analysis, but because they decide to show the data in such an easy way that puts at risk important information that should be captured. The use of spider plots is still a common practice that many people can easily understand, but the problem is that this analysis is so simple that it can mask sensory complexity.

It is worth noticing that there is a third type of analysis in the data science field, **predictive analysis**. This is a hot topic in the area that involves techniques such as data modeling, machine learning, AI, and deep learning. Instead of being focused on exploring or explaining the data, predictive analysis is concerned with making successful predictions, in ensuring that the predictions are accurate. Examples of this approach include face recognition and text-speech transcription. Eventually, some models can be studied to provide insights, but this is not always the case.

## 11.3 Audience Awareness

One of the most important things about being a successful consultant is Audience Awareness! No matter how good you, your team, and the service or product your company offers, if you fail to communicate with your target audience, the message will not get through. Knowing the audience, who the target people are is the cornerstone of any successful business. Knowing your audience makes you be better able to connect to with them.

In order to know your audience, you must gather some information about them beforehand, such as:

- **Background**: Do they have a sensory science background? Statistical background? Do they have experience in data science, including R language? Do they have experience in automated reporting dashboards, machine learning, etc? If so, are they juniors, specialists, or seniors?
- **Role**: What is the role your audience play in the project? Are they the decisions makers? Are they the final users of a dashboard, for instance?

- **Vocabulary**: Will your audience understand very technical terms or do you need to use simplified terms to convey the same message? This topic is closely related to the audience's background.
- **Expectations**: What is your audience expecting in a presentation or final report? A short summary of the project's outcomes? A detailed explanation of the statistical analysis including appendixces with further details? Recommendations for follow-up projects? Interpretation and conclusion of the analysis?

In general, according to the profile, the audience will likely fall into one of the three categories: **Technical Audience**, **Management (Decisions Makers)**, and **General Interest**. There is no magical formula on how to deal exactly with each of these types of audience, but in general, based on our experience, we must highlight that the key differences are in the focus, language, level of technical and detailed information you need to provide for each of those target public. In general, it tends to be necessary a higher level of details and technical information and a lower level of the big picture once you move from your general interest audience to the management and further down to the technical audience (Figure 11.1). We will further discuss the main differences between each audience above.

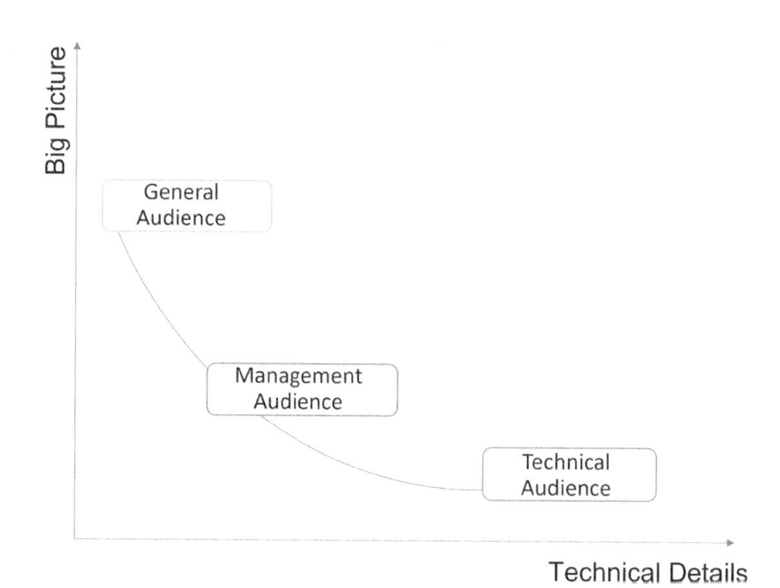


**FIGURE 11.1**
Trade-off curve based on the level of technical details and the big picture for each type of audience.

### 11.3.1 Technical Audience

The technical audience refers to the ones who will likely have a significant background and experience in or related to the field you are providing consulting service (e.g. Sensory Scientists, Statisticians, Data Analysts, Data Architecture, or your client's IT group). They are the team that will likely be working closely with you throughout the project development, at different stages. This type of audience is usually more exigent and/or engaged and because of their expertise in the field, will likely be expecting a presentation, report, or any other technical document in a higher level of details and with a more technical vocabulary; otherwise, you will sound that you are not an expert in the topic. This audience usually needs a lower sense of the big picture of the project, which means, that they are less interested in the details like the timelines, main outcomes, etc. But be aware, it is very important to still be able to distinguish between different technical audience (e.g. don't use sensory technical language to talk to the IT team).

### 11.3.2 Management

Although a person in a management position (e.g. Sensory Manager or Director and Principal/Senior Scientist) likely has a broad experience and background in the field, they tend to be more interested in the whole picture, which means timelines, progress of the project, potential issues, outcomes, applicability, next steps, etc. A person in a management position has many other projects and roles in a company and will not have time to be involved in the details. Instead, they likely designate a team (your technical audience) to be closely involved. In this case, you should be more concise in a meeting, presentation, or report, for instance. It is advised to keep a certain level of technical language, but it is better to present things in a simpler way and in a lower level of details than you would do for the technical audience. Additionally, the focus should be different, since as we mentioned, this audience is likely to be more interested in the whole picture instead of the specifics of the project.

Another distinct type of audience that falls into the management audience would be the executives, as a VP of Research & Regulatory, for instance. This public is not necessarily from the field and has even less time and/or background to absorb the specifics. The focus should be the same (whole picture) but with even fewer technical details. The approach and language of this audience tend to be closer to the general interest.

### 11.3.3 General Interest

The general audience usually refers to the ones that are likely the final users or are somehow related, contributed, or are interested in the project. In this way, this public is usually the least interested in the details and the most interested in the whole picture. The general audience usually refers to a larger group of people with different backgrounds and distinct levels of expertise, for instance, an R&D internship, a Chemistry Researcher, and a Senior Sensory Specialist can be all final users of a dashboard you developed. In this case, to make sure everyone follows you in a training (say), you must use less technical language and a lower level of details, otherwise you will lose part of your audience's attention. But at the same time, you may need to consider covering things that sound obvious to you; you ought to be careful about not skipping topics assuming that everyone knows about that or using certain terms and expressions considering that is evident for all. This can be the most challenging audience to deal with due to its diversity, but in a meeting, training session, or presentation, you should be very attentive and use your feeling to capture what is going on and maybe change your position to better connect with the audience.

A valuable tip shared by Cole Nussbaumer, in her book *Storytelling with data* (Knaflic, 2015), is to avoid general audiences, such as the technical and management team at the same time, or general audience such as "anyone related to the field that might be interested in the project". Having a broad audience will put you in a position where you can not communicate effectively to any of them as you would be if the audience was narrowed down.

- **Example**

We will use the PCA biplot from the biscuits sensory study shown in Chapter 10 (Figure 11.2) and point out the main differences in the approach according to the audience. As a quick reminder, 11 breakfast biscuits with varying contents of proteins and fibers were evaluated in this study. Products `P01` to `P09` are prototypes, product `P10` is a standard commercial biscuit without enrichment, and the 11th product (`Popt`) is an additionally optimized biscuit.

Let's picture a situation where the R&D team has been developing multiple trials for the biscuit formulation, changing the concentration/ratio of protein and fiber, with the objective to have a product with a sensory profile as close as possible to the commercial biscuit. For this exercise, your role as a consultant was to support the R&D team designing the study and conducting the analysis and ultimately analyzing and interpreting the results to make the final conclusion.

We will not go deep into the interpretation since it's not the focus of this example, but rather point out the approach we would recommend for each type of audience as shown in Figure 11.3.

Following those recommendations, your PCA for the management or general audience may look like Figure 11.4.

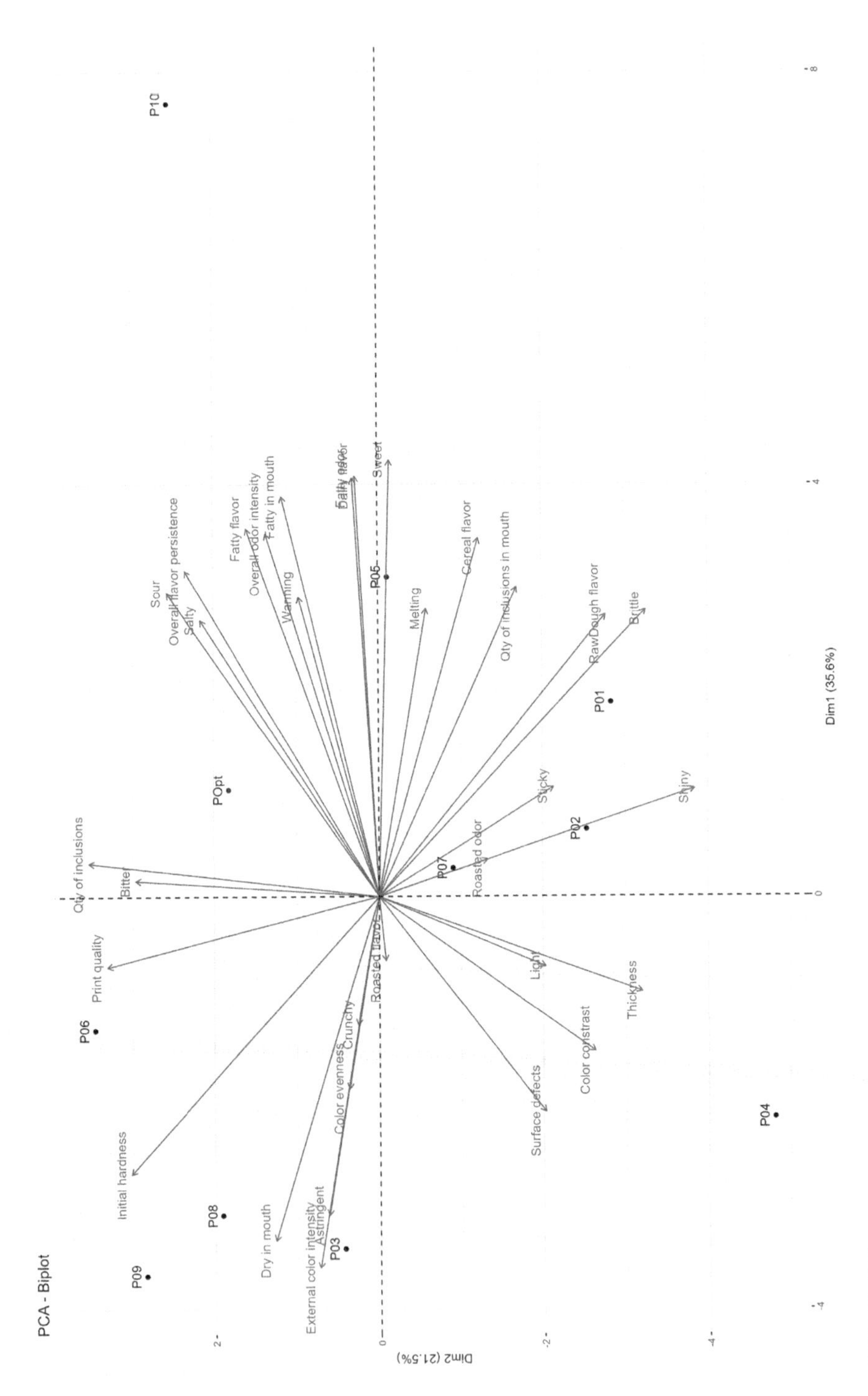


**FIGURE 11.2**
PCA biplot biscuit study.

| Topics | Technical Audience | Management Audience (Management / Executive) | General Audience |
|---|---|---|---|
| Context/Background Information | The objective of the study, including general and specific objectives<br><br>Information on the design of the experiment<br><br>How samples were selected or screened for the study, including a detailed description of each sample<br><br>Details about the panel such as number of panelists, level of training, and how panelists were trained for the attributes evaluated<br><br>Details about the test, how attributes were defined, and how and where samples were evaluated | The level of information is somewhere between the Technical Audience and General Audience.<br><br>In the case of the management audience, it is recommended some level of details you deem important and the use of technical language. Also, be prepared to show or discuss a deeper level of information as requested<br><br>In the case of an executive audience, the level of information tends to be closer to the general audience | An overall and succinct description of the general objective of the study, samples tested, and type of test performed |
| Data Exploration | It may be important to show/mention the panel performance evaluation<br><br>It is important to show the PCA diagnostics, including the Scree Plot and the plots displaying the contribution of each attribute in each dimension | | It is not recommended to share the data exploration and analysis done before the PCA.<br><br>The general audience does not need to understand details such as why two components were chosen for the PCA |
| Final Results (PCA) | The PCA can be shown in its traditional format without much editing as shown in the picture above<br><br>Provide a detailed interpretation, including not only the optimized sample, but all trials evaluated | | Simple and clear explanation about PCA<br><br>Remove technical details such as the PC variance explanation<br><br>Show the PCA in a format that makes the interpretation and visualization easier and more obvious, some examples include:<br><br>- Use pictures instead of sample code<br><br>- Use visual resources like circles to delimit groups of similar samples and attributes more relevant<br><br>- Use different colors, letter sizes, or other visual resources to highlight important information, like an attribute or a sample |
| Conclusion and Recommendations | The conclusion can be more detailed, and recommendations can be done for further studies | | Very clear and succinct conclusion or recommendation.<br><br>Example: The optimized formulation is close to the commercial but needs to be improved for some attribute |

**FIGURE 11.3**
Exemplification of the approach for each type of audience.

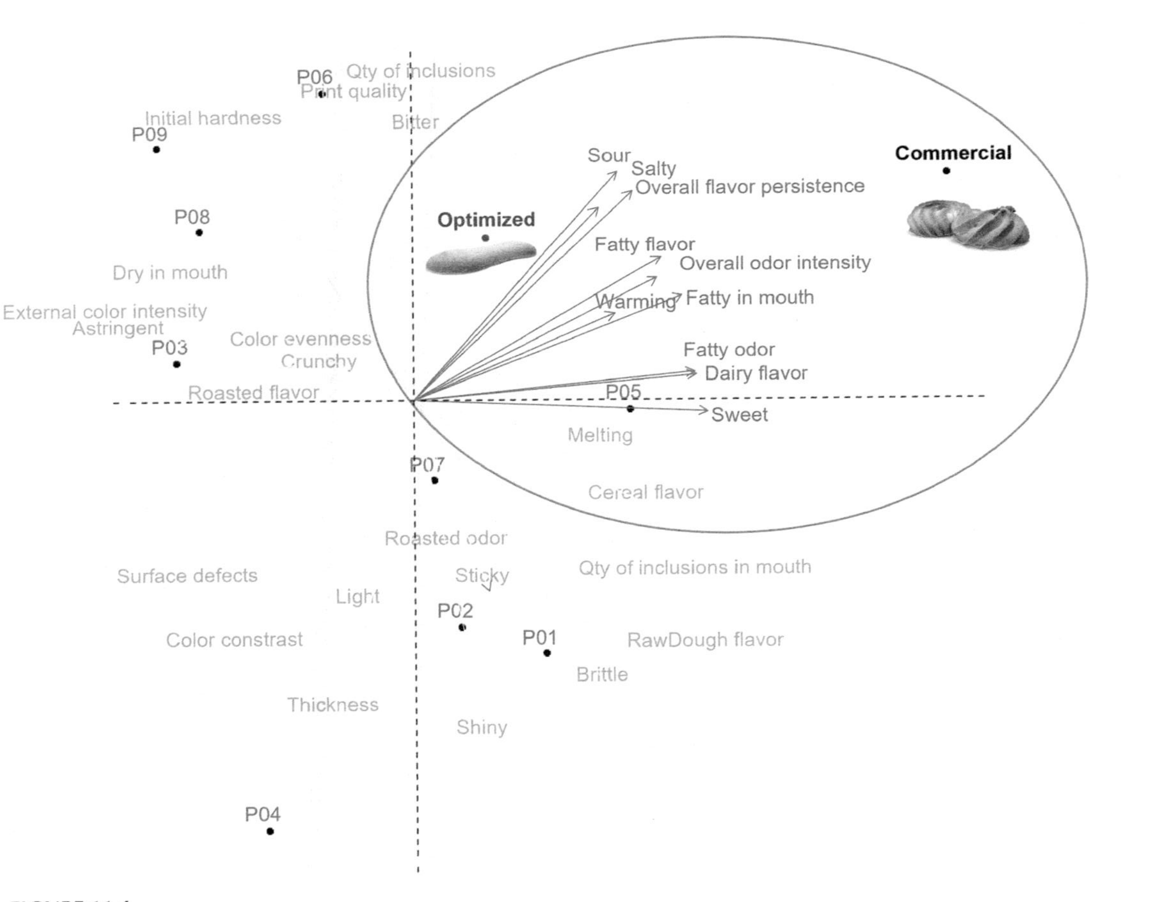


**FIGURE 11.4**
PCA biplot biscuit study modified.

Note that the PCA was simplified for a more straightforward understanding:

- the PCA variance explanation and grid lines were removed,
- attributes were slightly moved to avoid overlap,
- the samples and attributes with lower interest were given a lighter color,
- the attributes and samples we want our audience to focus their attention were given a different, stronger color,
- pictures and a more appealing description were used instead of the samples codes
- some strategies as to circle the important area/group of samples and attributes helps the audience to focus on what we deem most important to extract from this analysis were adopted.

In this example, the idea may be to highlight the audience that the optimized formulation is in fact closer to the commercial one, and to increase even more this similarity, some attributes, like sour, salty, overall flavor persistence, fatty flavor, and fatty in mouth, have to be increased.

## 11.4 Methods to Communicate

How will you communicate to your audience? Are you going to deliver a live presentation? Are you going to present a proposal in a live meeting? Or will the communication be a written document you will send through email?

What is the format you will be using to communicate? Word, Excel, or PowerPoint? Are you going to send the document in PDF format? Are you going to present a dashboard? Are you going to share R Scripts?

As we will discuss in more detail later, the way and the format you use to communicate to your clients or audience have a huge impact on successful communication, and you should be well aware of that!

### 11.4.1 Consider the Mechanism

You should be aware that the primary method or mechanism you use to communicate strongly affect the way your audience effectively gets the information and so you should tailor it accordingly. One of the most important aspects is related to the amount of control you have over the audience, how they get the information, and hence the level of details needed (Knaflic, 2015).

In a live presentation, for instance, you are in full control. You can answer questions your audience may have, you can slow down and go into a particular detail you deem important, or you can speed up over repetitive, obvious, or not-so-important topics. In short words, you are the expert there and so, you can easily provide effective communication, and because of that, you don't need to overcrowd your slides or any other document and divert or lose your audience's attention with unnecessary information. You can, for instance, just show a plot or graph and a very simple interpretation or bullet points because you are covering vocally the nuances and details about that.

In the case the communication is done through a written document in a non-live situation, you have much less control over your audience, on how they will take the information, and on whether they will get the main point. In this situation, you need to be more careful and likely provide a higher level of details to answer or clarify potential questions or doubts your audience may have. In this situation, showing a plot or graph and just a very simple interpretation or bullet points is likely not enough.

It can be a great idea to merge those two formats, when possible, where you can give time to the audience to consume the information on their own for a while and give the topic thought and a moment where you can discuss it in a live situation, not in this order necessarily. So, for example, let's pretend you have to present a proposal for a client. Instead of sending a dense document to explain all the details and just wait for the client's response, you can make a more concise document, easier to go through if you have a live moment with the client. You can for instance present the proposal initially in a live meeting, where you cover in general all the important topics and details and then send the written document to the client.

### 11.4.2 Pick the Correct Format

The second point of this topic on the method of communication is related to the correct format to pick. There are certainly many ways for you to communicate with your client – word, excel, or PowerPoint whether in pdf format or not, dashboard, or even scripts – but surely one is the most suitable. Again, there is no universal answer for the best format to pick since it may vary according to clients' requests and the type of project you are dealing with. But there is one thing you should always follow, unless strictly necessary, do not share documents in an editable format. You may use Word to write proposals or final reports, Excel for plots or tables, and PowerPoint for live presentations, and that is totally fine, but never share that in the editable format. We always recommend saving in pdf format to share with your client or audience and this is because of two simple reasons. First, the pdf format cannot be modified! You definitely do not want to take the risks of others changing your document, which can lead to misunderstanding, putting you in a delicate situation. Second, the pdf format preserves document formatting which means that it retains the intended format if the file is viewed online or printed. In short words, it is very unprofessional to share documents in editable format.

You may be wondering, so in what type of situation would you share an editable format? When would you share a document in Word, PowerPoint, or Excel? In the situation where you are working with a partner for instance. So, for example, a project that you are working on involves multiple partners in a common report or presentation. In this case, it may be convenient to share the Word or PowerPoint document for each of the partners to include their inputs. After the document is ready, make sure you carefully review the formatting and save it in pdf before sending it out to the client.

There are two other formats that may be common in the sensory data science field, which are dashboards and scripts. If you are developing a graphical user interface for your client, you need to deploy the dashboard at some point to a server for your client to be able to access it. The deployment can be done in two ways: web-based, as a simple client web page, or locally, as a locally installed desktop application. The choice should be based on the client's preference.

The last method of communication that is fairly common in the field is R or any other programming language script. It is very common that the client requests the scripts used for a specific project, the text file containing the set of commands and comments you used, for instance, to develop an automated analysis reporting dashboard. You can share the repository where the scripts are hosted, or you can zip the scripts and share them with your clients. The details should be discussed with the client's IT team since each company has a particular preference. As the scripts should always be available under the client's requests, you should be careful to not display sensitive or confidential information by reusing codes or throughout the comments.

## 11.5 Storytelling

There are basically two ways to communicate with our audience, the first is called conventional rhetoric. A PowerPoint full of facts, filled with bullet points and statistics with a presenter with a formal and memorized speech and using the same voice tone, would be the best way to illustrate the conventional rhetoric style. This way to communicate, which drove the businesses of the past, has a more analytical approach, where statistics, charts, metrics would be dumped on the audience and left to them to digest. There is no need to say that this approach is completely outdated, it clearly fails to stimulate the audience's attention or evoke their energy or emotions. The second way to communicate, which is the last topic we want to cover in this chapter and also happens to be a critical skill for any successful consultant, is through storytelling! Storytelling is something that we all know, from an early age we were introduced to the notion of narrative structure, which means a clear beginning, middle, and end. The ability of one going throughout this structure to tell us a story is what makes a book, play, or movie grab our attention and evoke our emotional responses, is what makes it interesting! In short words, storytelling is one of the most powerful and effective ways to attract people's attention because we were taught to communicate with stories throughout history. This universal language that everyone can understand has the power to truly engage your audience because it translates abstract facts, numbers, and plots into compelling pictures; it inspires, motivates, and drives actions because it taps into people's emotions.

As described in the book *Once Upon an Innovation*, by Jean Storlie and Mimi Sherlock (Storlie and Sherlock, 2020), the left side of our brain is linked to more logical and analytical thinking, including data processing, number handling, and statistical interpretations. The right side is linked to expression, emotional intelligence, and imagination, and in our context, will be the part of our brain that will capture the big picture, that will turn data and facts into possibilities and innovative ideas. If you as a consultant overwhelms your audience with analytics you will reduce their capacity for big picture thinking, you will shut down their capability to generate novel ideas and solutions. We are not saying that numbers, plots, and facts are not important, but that they should be presented in a story narrative format, in a way that will be able to light up the right side of the brain, and this stimulation of both paths is what trigger unexpected and novel solutions, inspire support and drive changes.

In a real situation, you as a consultant have many pieces of information that you have collected throughout the journey with a client, from the very first communication until the end of a project. You have valuable information about your client company's situation, challenges and issues, needs, expectations, potential solutions and/or failed attempts, and final outcomes. Storytelling is the master of tying all together and articulating it into the context of a story in a creative way to engage and persuade your audience. A good story allows you to successfully connect with your audience, it makes your audience understand, reflect, and act in a way that plots, numbers, and facts altogether simply can't.

You may be wondering. How exactly should I construct a story? What should be covered in each piece of the narrative? We will provide here a summary of the pieces of a good story and the specifics based on our experience and also on the books *Storytelling with Data* by Cole Nussbaumer (Knaflic, 2015) and *Beyond Bullet Points* by Cliff Atkinson (Atkinson, 2018). Both books dedicate a good part to Storytelling, making them a great resource on this topic.

### 11.5.1 The Beginning (Context)

The key piece of any story is the context, the description of the situation, and surrounding details. This first step is the moment to set up the essential information or background on the topic you will be covering to get everyone on common ground. You should initially spend time to make sure your audience clearly understands the context, why this is important or necessary, and why they are there before diving into actions or results. Subsequently, you will raise the challenges or problems and propose some recommended solutions. It is at this very first step that you will first grab your audience's attention. If you fail at this moment, it is very unlikely that you will recover their interest in the subsequent steps.

For live presentations, it is strongly recommended to use the first few minutes to be an icebreaker to make everyone feel more comfortable and create a more friendly environment. In order to do so, you can start introducing yourself in case you have not met everyone yet, you can have a conversation and talk about the latest news, ask about how they and their families are doing, etc. The second piece of advice is to start the presentation by stating bullets of the main points that will be covered, so your audience will have awareness of what you will be talking about.

### 11.5.2 The Middle (Action and Impact)

Now is when you get to the crux of your story, it is at this moment that you will explain your solutions or actions and highlight the impacts. You will continue it in a way you will convince your audience of the solution you are proposing or make them clearly understand, agree, and be excited about the outcomes and possibilities of a solution you worked on. You should be careful to retain your audience's attention addressing how they can be part and/or benefit from the solution you are referring to. In the case of a live presentation, pose always confidently, show enthusiasm about what you are talking about, and watch out for hidden clues, try to constantly catch your audience's response/feedback through their expressions and body language.

The content to build out your story at this moment is very dependent on the context of the situation, but from a consultant perspective, it will be likely the moment you will further develop the situation or problem covering relevant information, show some data to illustrate the situation, discuss potential solutions to address a particular topic or present the outcomes of your project.

### 11.5.3 The End (Conclusion)

This is the moment you close your story; it is when you should tie it back to the beginning to somehow recap the problem, highlight the basic idea, and conclude the story. You should finish your presentation in an impactful way, reemphasizing and repeating your main point, what you want to stick deeply in your audience's mind. Once more, the content at the end of the story can be somehow dependent on the context of the situation, but in a consulting world, it would likely include a conclusion of the topic and also next steps and further recommendations.

## 11.6 Reformulate

Something important to keep in mind is the follow-up process after a report is sent or a presentation is delivered. The ability to receive feedback and reformulate is undoubtedly a very important and sometimes challenging skill that consultants seeking success should be aware of. It may be challenging since some consultants can be reluctant to feedback because of a misconception that they are the experts in the field and hence, their approach is the best. So, one of the most important rules, regardless of the expertise and knowledge you have in the field, is to be humble! Consultants must understand the idea that: 1) you need to make your client pleased unless you have a strong reason not to do so like ethical reasons or statistical rules and 2) your point of view can be biased over time and your client's request may indeed improve the clarity of an outcome for instance. Or, you can simply be wrong, miss something, and have not taken the best approach. It happens! So, be open to feedback and be prepared to reformulate!

Sometimes the client feedback is something very minor, to adjust the scale of a plot, match the color with the company's palette, or change the type of plot. In other cases, the feedback will demand a bit more time. It is common that the way you deemed best to present the outcomes is not that clear from your client's view or the set of data or plots you selected did not convey the message you were expecting or in an extreme situation, your client does not agree or ask you to redo an experiment or procedure. In this case, you will need to dedicate more time to address your client's request.

Regardless of the situation, you should be motivated and be open-minded to your client's feedback and afterward carefully work on that to tackle it all at once. You definitely want to avoid a situation where your report or presentation be back and forth with your client. It is recommended that you make all possible changes and prepare a convincing explanation for the things that you strongly do not agree with or have a solid reason not to do so. Ideally, you should get back to your client as soon as possible highlighting the changes that were made and explaining the ones not addressed.

# 12

# *Machine Learning*

Artificial Intelligence (AI) and Machine Learning (ML) have gained a lot of attention in recent years. With the increase in data availability, data storage, and computing power, many techniques that were just *dreams* back then are now easily accessible and used. And of course, the sensory and consumer science field is not an exception to this rule as we start seeing more and more ML applications although in our case, we do not have *Big Data* per se, but we do have *diverse data*! For many of us, AI and ML seems to be a broad and complex topic. This assertion is true, and in fact it would deserve a whole book dedicated just to it. However, our intention in this chapter is to introduce and demystify the concept of ML, by: (1) explaining the differences between supervised and unsupervised ML models, (2) proving that you were already doing it long ago, perhaps without knowing, (3) extending it to more advanced techniques, (4) highlighting its main applications in the field. To do so, some basic code and steps will be provided to the reader to get familiar with such approach. Throughout this chapter, some more specialized resources are provided for those who have the courage and motivation to dig deeper into this topic.

## 12.1 Introduction

Machine Learning is currently a hot topic in the sensory and consumer science field. It is one of the most game-changing technological advancements to support consumer packaged goods companies in the development of new products, playing a considerable role in speeding up the R&D process (and at the same time reducing the costs). In today's fast-moving and increasingly competitive corporate world, companies that are embracing, adopting, and opening their minds to digital transformation and artificial intelligence (AI), moving toward the age of automation, are not one but many steps ahead of their competitors.

DOI: 10.1201/9781003028611-12

Machine Learning (ML) is a branch of AI, which is based on the idea that systems can learn from data and that has the capability to evolve. Generally speaking, ML refers to various programming techniques that are able to process large amounts of data and extract useful information from it. It refers to data analysis methods that build intelligent algorithms that can automatically improve through the experience gained from the data and can identify patterns or make decisions with minimal human intervention, without being explicitly programmed. ML focuses on using data and algorithms to mimic the way humans learn, gradually improving their accuracy.

Defining the objectives or the situation where ML would bring value is the very first step of the process. Once that is clear, the next step is to collect data or dig into historical data sets to understand what information is available and/or has to be obtained. The data vary according to the situation, but it may refer to product composition or formulation, instrumental measurements (e.g. pH, color, rheology, GC-MS, etc.), sensory attributes (e.g. creaminess, sweetness, bitterness, texture, consistency, etc.), consumer behavior (e.g. consumption frequency, use situation, dietary constraints, etc.), demographics (e.g. age, gender, size of household, etc.), and consumer responses (e.g. liking, CATA questions, JAR questions, etc.) just to name a few.

First of all, it should be stressed that the size of the data set and its quality are very important as they impact directly the model's robustness. Here are general recommendations (to be adapted to each situation, data type, and objectives):

- The higher the number of statistical units the better, with 12–15 being the minimum recommended when statistical units correspond to samples.
- The number of measurements (instrumental, sensory, and/or consumer measurements) and the number of consumers evaluating the products are also very relevant to the model's quality. In practice, a minimum of 100 participants is usually recommended for consumer tests, which deemed sufficient to apply ML (although here again, the more the better). For data quality, the variability of the samples is one of the most important aspects (besides the standardization of data collection). The larger the variability between samples, the broader the space the model covers. Additionally, it is strongly recommended to capture the consumers' individual differences, not only through demographic information but also through perception (including rapid sensory description methods, Just About Right (JAR) or Ideal Profile Method (IPM)). Eventually, within-subject design (i.e. sequential monadic design) provides better quality models as they allow accounting for individual response patterns.

## 12.2 Introduction of the Data

For this section, we use the wine data set from the {rattle}.[1] This data set consists of the results of a chemical analysis of wines grown in a specific area of Italy. In total, the results of 13 chemical analyses (e.g. alcohol, malic acid, color intensity, phenols, etc.) are provided for 178 samples that represent three types of wines.

```
library(tidyverse)
library(rattle)

wine <- rattle::wine %>%
  as_tibble()
```

## 12.3 Machine Learning Methods

The notion of ML is vast, as it covers a large variety of analyses. In fact, ML algorithms are often classified based on the goals of their analysis. Three main groups are often considered:

- Unsupervised Learning:

Unsupervised ML aims at finding structure within the data. Input are unlabeled data, meaning that no *output* values are yet known. In this case, the algorithms operate independently from any information about the data to find patterns and trends. For instance, this is achieved by learning from the data distribution the features that distinguish between statistical entities using similarity and dissimilarity measurements. Such ability to discover *unknown* patterns in the data makes such algorithms ideal for exploratory analysis. In Sensory and Consumer Science, the best known Unsupervised ML techniques are Principal Component Analysis (PCA) for dimensionality reduction and hierarchical cluster analysis (e.g. for consumer segmentation).

- Supervised Learning:

Supervised ML is arguably the most popular type of ML: When people talk about ML, they often refer to Supervised techniques. Supervised ML

[1] https://rdrr.io/cran/rattle.data/man/wine.html

takes labeled data as input, meaning that the statistical entities are defined by one or more output variables. The aim of the algorithm is then to a find a mapping function that connects the input variables with those output variables. Ultimately, the ML model aims to explain output variables using the input variables. A common situation requiring Supervised ML in Sensory and Consumer Science consists of predicting consumer responses (e.g. liking) using sensory descriptions, analytic data, demographics, or any other information. ML models provide insights on how to improve product performance and allow predicting consumer responses of new prototypes or products. Another common situation is to use Supervised ML to predict the sensory profile of products using formulation data (i.e. ingredients and process parameters).

- Semi-supervised Learning:

Semi-supervised ML is not an ML approach per se. Instead, it is a combination of both Unsupervised and Supervised approaches. It first aims to create an output variable using Unsupervised techniques and then to explain or use this output variable using other information through Supervised ML. A good example of semi-supervised approach consists of defining clusters of consumers based on liking (unsupervised) and to characterize these clusters using demographic data using decision trees for instance (supervised). External Preference Mapping is another example since it first reduces the dimensionality of the sensory data through PCA (unsupervised) and then uses these dimensions to explain the consumers' liking scores using regressions (supervised).

---

A forth type of Machine Learning is called *Reinforcement Learning* that relies on feedback provided to the machine. It is a technique that enables an agent to learn through trial and error from its own actions and experiences. Reinforcement Learning is commonly used in some tech applications (e.g. gaming and robotics), especially when large data sets are available. Such approach has little reach in sensory and consumer science at the moment. Therefore, we are not going to develop it further here.

---

## 12.4 Unsupervised Machine Learning

In sensory and consumer science, unsupervised learning models are mainly used for Dimensionality Reduction and for Clustering.

### 12.4.1 Dimensionality Reduction

Dimensionality reduction is a technique used to transform a high dimensional space into a lower dimensional space that still retains as much information as possible. In practice, the original high-dimensional space involves many variables that are correlated with each other, but that could be summarized by *latent* variables or *principal components*, which are orthogonal to each other.[2]

Most frequently, dimensionality reduction is performed for the following reasons:

- Summarizing data (and removing redundant features);
- 2D or 3D visualization of the data (most important information);
- Finding latent variables and untangling initial variables;
- Preprocessing data to then reduce training time and computational resources;
- Improving ML algorithms accuracy by removing the lower dimensions (the one containing less information) often considered as *noise*;
- Avoiding problems of over-fitting.

Some of these approaches were presented earlier in this book, in particular in Chapter 10. However, there are numerous dimensionality reduction methods that can be used depending on the data at hand. The most common and well-known methods used in the sensory and consumer science are the ones that apply linear transformations, including Principal Components Analysis (PCA), Factor Analysis (FA), and derivatives such as (Multiple) Correspondence Analysis, Multiple Factor Analysis, etc.

Let's apply this technique to the `wine` data. To get familiar with the data, we can first visualize the information on a 2D plot and then reduce the data set to the first two dimensions only.

Since the different variables represent analytical measures that are defined using different scales, a standardized PCA is performed. This is the default option in `PCA()` from {`FactoMineR`}(`scale.unit=TRUE`):

```
library(FactoMineR)

res_pca <- PCA(wine, quali.sup=1, scale.unit=TRUE, graph=FALSE)
```

[2] When all the *principal components* are considered, none of the information present in the raw data is lost and their representation is simply shifted from an unstructured high-dimensional space to a structured low*er*-dimensional space.

The results of the PCA (Figure 12.1) can be visualized using {`factoextra`}:

```
library (factoextra)

fviz_pca_biplot(res_pca, repel=TRUE, label="var",
                col.var="red", col.ind="black")
```

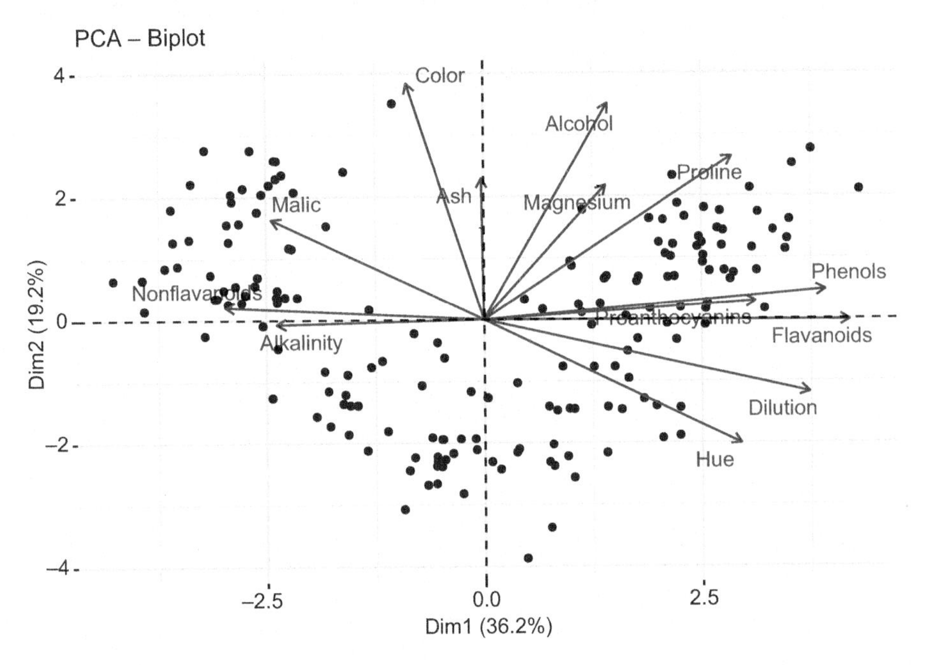


**FIGURE 12.1**
PCA Biplot.

The first plane of the PCA suggests that there are three distinct groups of wines. Let's define them mathematically using cluster analysis. For this process, we propose to reduce the full data to its first two components only. This approach is used here to illustrate how PCA can be used as a preprocessing step. Additionally, such preprocessing can help detecting clearer patterns in the data, as we will see in Section 12.4.2.

```
wine_reduced <- as_tibble(res_pca$ind$coord[,1:2])
```

---

Note that such use of PCA as a preprocessing step was already done earlier in Section 10.4.4 when the sensory space was reduced to its first two dimensions before performing the external preference mapping.

---

### 12.4.2 Clustering

Clustering is a technique used when dealing with high-dimensional data to discover groups of observations that are similar to each other (or different from each other). In other words, it is a method that groups unlabeled data based on their similarities and differences in a way that objects with strong similarities are grouped together and are separated from objects to whom they have little to no similarities.

Again, a very common application in S&C Science is to segment consumers based on a variety of factors such as shopping or usage behavior, attitudes, interests, and preferences. As consumers being associated in the same market segment tend to respond similarly, segmentation is a key strategy for companies to better understand their consumers and tailor effectively their products or marketing approaches for the different target groups. Similarly, it is also used to classify products in homogeneous groups based on their analytical, sensory, and/or consumer description in order to analyze the product offer and identify different segments on the market.

There are many clustering approaches and algorithms. They can be categorized into different types including *exclusive* (e.g. k-means), *hierarchical* (see Section 10.4.2 for an example), and *probabilistic* (e.g. Gaussian Mixture Model). The first two are most widely used and well known in the sensory field.

Although agglomerative hierarchical clustering (HAC) is more common in sensory and consumer research, an example illustrating such approach was already provided in Section 10.4.2. For that reason, here we propose to present another approach using *k-means*. K-means clustering is a popular unsupervised machine learning algorithm for partitioning a given data set in a way that the total intra-cluster variation is minimized. Both approaches (HAC and k-means), however, differ in their ways of forming clusters. For instance, the algorithm of k-means requires the user to prespecify the number of clusters to be created, whereas HAC produces a tree (called dendrogram) which helps visualizing the data hierarchical structure and deciding on the optimal number of clusters. Detailed information about clustering methods and analysis can be found in the book *Practical Guide to Cluster Analysis in R: Unsupervised Machine Learning* by Alboukadel Kassambara (Kassambara, 2017a).

In order to cluster our wines, let's start with defining the optimal number of clusters (k) to consider. This can be done using `fviz_nbclust()` from {factoextra}. This function creates a graph that represents the variance within the clusters (Figure 12.2). In this representation, the bend (also called *elbow*) indicates the optimal number of clusters; any additional cluster beyond that point has less value.

```
fviz_nbclust(wine_reduced, kmeans, method = "wss")
```

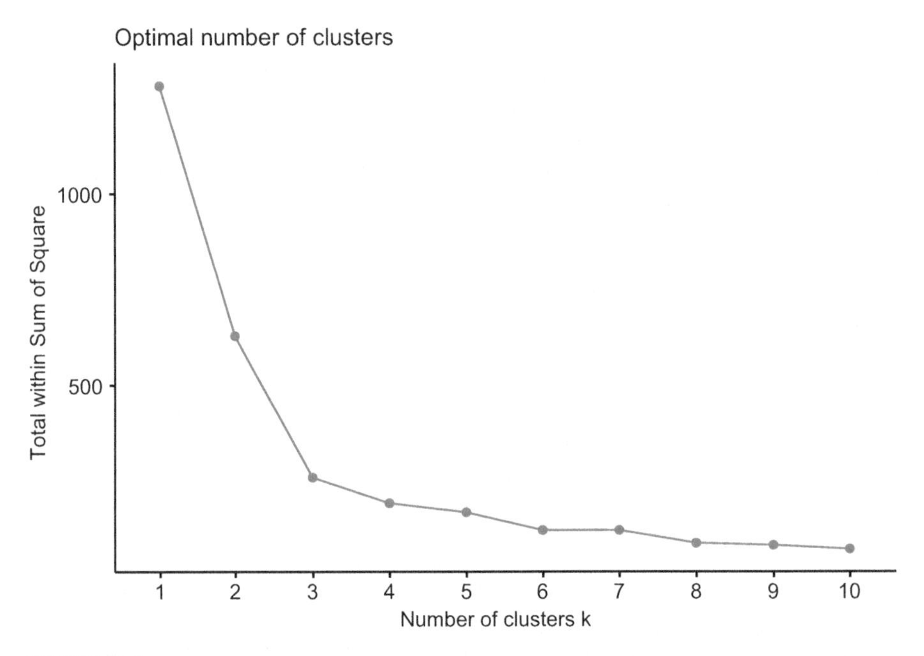


**FIGURE 12.2**
Optimal number of clusters.

Here, the optimal solution consists of defining three clusters.

Next, the k-means algorithm starts with randomly selecting k (here 3) centroids. In order to reproduce our results (despite the randomness), we propose to initially set a seed (through `set.seed ()`).[3] Otherwise, it is recommended to set within `kmeans()` a number of random sets, that is, the number of times (here 20) R will try different random starting assignments. Increasing this number yields more stable results.

[3] `set.seed()` fixes random selections so that they can easily be reproduced.

```
set.seed(123)
wine_kmeans <- kmeans(wine_reduced, centers=3, nstart=20)
```

Finally, the results (Figure 12.3) can be visualized using `fviz_cluster()` from {`factoextra`}:

```
fviz_cluster(list(data=wine_reduced, cluster=wine_kmeans$cluster),
            ellipse.type="norm", geom="point", stand=FALSE)
```

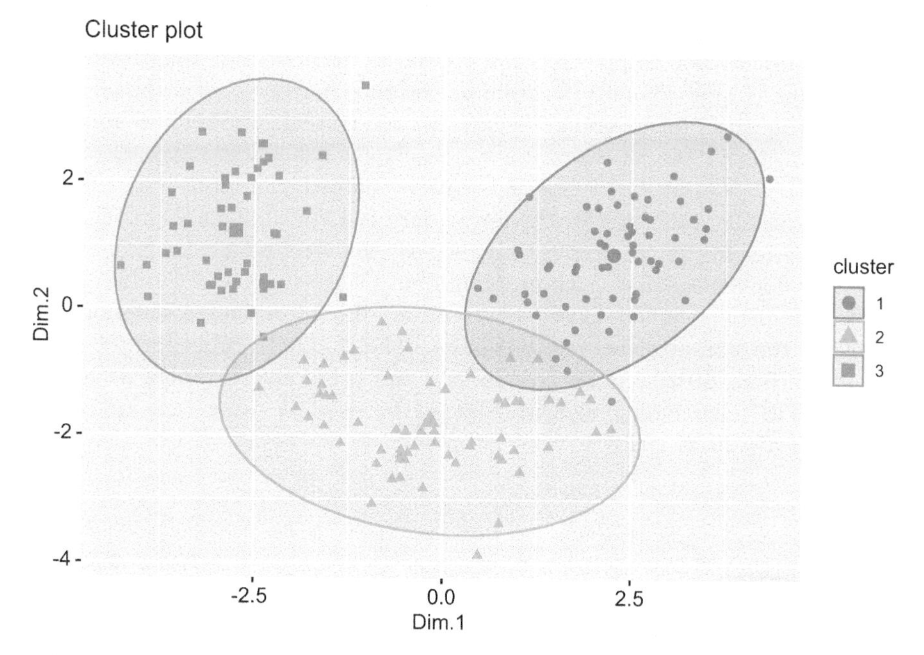


**FIGURE 12.3**
Cluster plot.

In the resulting plot, the three clusters are clearly separated, as can be seen by their little to no overlap.

---

An interesting suggestion is to run the same analysis on the full data (here, we limited it to the first two dimensions of the PCA) and to compare the results.

---

When applying such clustering techniques, one may sometimes encounter very atypical variables that could be deemed noisy. This may occur when analyzing consumer hedonic data and finding few consumers who have an atypical response pattern. This may create enough "noise" in the data to blur the main data structure and affect the outcome of the k-means algorithm. To solve this problem, Vigneau et al. (2016) have developed strategies to segment consumers while setting aside atypical or irrelevant consumers. This can be achieved by using either a *noise cluster* where to dump "irrelevant" consumers or a sparse latent variable (Sparse LV) strategy. Both strategies have been implemented in the {`ClustVarLV`} package and can be selected using "kplusone" or "sparselv" parameters in the `CLV_kmeans()` function.

It should be noted that, in sensory and consumer science, it is increasingly more frequent to aim to cluster panelists, based not only on one variable (i.e. their liking for a series of products) but also on sets of several variables (i.e. their description of products for a series of attributes or their responses to a full online questionnaire) or even on distance matrices (as obtained from a free sorting task, for example). Segmentation strategies would also apply to such data.

For instance, Cariou and Wilderjans have developed an approach for clustering around latent variables for three-way data (CLV3W). This approach that is implemented in {`ClustVarLV`} could be used to detect panel disagreement in three-way conventional sensory profiling data (Wilderjans and Cariou, 2016) or to segment consumers based on multi-attribute product evaluation, by removing the non-negativity constraint to the `CLV3W()` procedure (Cariou and Wilderjans, 2018). As for free sorting and projective mapping data, the CLUSTATIS partitioning algorithm could be applied using {`ClustBlock`} (Llobell et al., 2019, 2020).

## 12.5 Supervised Learning

There are many ways to carry out Supervised ML, which again would require an entire book dedicated just to it. In this section, we will introduce you to the basics, which should give you a nice kick-start for your own analysis. For those who want to learn more on this topic, we recommend reading *Hands-On Machine Learning with R*[4] by Bradley Boehmke and Brandon and *Tidy Modeling with R*[5] by Max Kuhn and Sylvia Silge for more in-depth information.

[4] https://bradleyboehmke.github.io/HOML
[5] https://www.tmwr.org/

### 12.5.1 Workflow

In sensory and consumer science, supervised learning is commonly carried out using a regression type of analysis, where for instance consumer ratings are used as output (target) and product information (i.e. sensory profiles and/or analytical measurement) are used as input. The goal of the analysis is then to explain (and sometime predict) the (say) liking scores using the sensory information about the products.

To do so, models are initially trained using a subset of the data (called *training set*). Once obtained, the model is then tested and validated on another part of the data (called *test set* and *validation set*).[6] Once this process is done, the model can be continuously improved, discovering new patterns and relationships as it trains itself using new data sets.

### 12.5.2 Regression

Regression methods approximate the target variable[7] with (usually linear) a combination of predictor variables. There are many regression algorithms varying by type of data they can handle, type of target variable, and additional aspects such as the ability to perform dimensionality reduction. The most relevant methods for sensory and consumer science will be presented here.

- **Linear regression**:

The simplest and most popular variant is linear regression in which a continuous target variable is approximated as linear combination of predictors in a way that the sum of squares of the errors (SSE) is minimized. It can be, for example, used to predict consumer liking of a product based on its sensory profile, but the user has to keep in mind that linear regression can in some cases return predicted values outside the reasonable range of target values. This can be addressed by capping the predictions to a desired range. Functions in R to apply linear regression are: `lm()` and `glm()` or `parsnip::linear_reg() %>% parsnip::set_engine("lm")` when using the {`tidymodels`} workflow.

- **Logistic regression**:

Logistic regression is an algorithm which – by use of logistic transformation – allows to apply the same approach as linear regression to cases with binary

[6] It is common practice to split the data so that the model is built by using 70% of the data, the remaining 30% being used for testing and validation. It is however important to use separate data for these steps to avoid any over-fitting.

[7] This is a bit of simplification since in some cases, it is some transformations of combination of predictors that approximate the target variable, as in logistic regression for example.

target variables. It can be used in R with `glm(family = "binomial")` or `parsnip::logistic_reg() %>% parsnip::set_engine("glm")` when using the {`tidymodels`} workflow.

- **Penalized regression**:

Often, the data used for modeling contain a lot of (highly correlated) predictor variables. In such cases of multicolinearity, linear and/or logistic regression may become unstable and produce unreasonable results. This can be addressed through the use of so-called penalized regression. Instead of minimizing pure error term, the algorithm minimizes both the error and the regression coefficients at the same time. This leads to more stable predictions.

There are three variations of penalized regression and all of them can be accessed via `glmnet::glmnet()` ($\beta$ is set of regression coefficients and $\lambda$ is a parameter to be set by user or determined from cross-validation):

- Ridge regression (L2 penalty) minimizes $SSE + \lambda \sum |\beta|^2$ and drives the coefficients to smaller values;
- Lasso regression (L1 penalty) minimizes $SSE + \lambda \sum |\beta|$ and forces some of the coefficients to vanish, which allows some variable selection;
- Elastic-net regression is a combination of the two previous variants $SSE + \lambda_1 \sum |\beta| + \lambda_2 \sum |\beta|^2$.

Penalized regression can also be ran in the {`tidymodels`} workflow using `parsnip::linear_reg() %>% parsnip::set_engine("glmnet")`.

- **MARS**:

One limitation of all above-mentioned methods is that they assume linear relationship between the predictor and the target variables. Multivariate adaptive regression spline (MARS) addresses this *issue* by modeling non-linear relationship with piece wise linear function. This gives a nice balance between simplicity and ability to fit complex data, for example, $\Lambda$-shaped once where there is a maximal point from which function decreases in both directions. In R, this model can be accessed via `earth::earth()` function.

- **PLS**:

In case of single and multiple target variables, partial least squares (PLS) regression can be applied. Similarly to PCA, PLS looks for components that maximizes the explained variance of the predictors, while simultaneously maximizing their correlation to the target variables. PLS can be applied with `lm()` by specifying multiple targets or in the {`tidymodels`} workflow with `plsmod::pls() %>% parsnip::set_engine("mixOmics")`.

### 12.5.3 Other Common Supervised ML Algorithms

Additional Supervised ML techniques include:

- **K-nearest neighbors**

A very simple, yet useful and robust algorithm that works for both numeric and nominal target variables is K-nearest neighbors. The idea is that for every new observation to predict, the algorithms find K closest points in the training set and use either their mean value (for numeric targets) or the most frequent value (for nominal targets) as prediction. This algorithm can be used with `kknn::kknn()` function or in the {`tidymodels`} workflow with `parsnip::nearest_neighbor() %>% parsnip::set_engine("kknn")`.

- **Decision trees**

Decision tree algorithms model the data by splitting the training set into smaller subsets in a way that each split is done by a predictor variable so that it maximizes the difference in target variable between the subsets. One important advantage of decision trees is that they can model complex relationships and interactions between predictors. To use decision tree in R, `rpart::rpart()` or in the {`tidymodels`} workflow `parsnip::decision_tree() %>% parsnip::set_engine("rpart")` can be used.

- **Black boxes**

The black boxes algorithm includes models for which the structure is too complex to directly interpret relationship between predictor variables and a value predicted by the model. The advantage of such models is their ability to model more complicated data than in case of interpretable models, but they have a greater risk of overfitting. Also, the lack of clear interpretation may not be acceptable in some business specific use cases.

- **Random forests**

A random forest is a set of decision trees, each one trained on a random subset of observations and/or predictors. The final prediction is then obtained by averaging the individual trees' predictions. By increasing the number of trees, we also increase the precision of the results. The random forest algorithm hence minimizes some of the limitations of a decision tree algorithm, by for instance reducing the risks of overfitting and increasing its precision.

## 12.6 Practical Guide to Supervised Machine Learning

Now that we have a general idea of the purpose of Supervised ML approach, let's build a simple machine learning model in the context of a sensory and consumer study. But before doing that, let's introduce the {tidymodels} framework.

### 12.6.1 Introduction to the {tidymodels} Framework

R contains many fantastic systems for building machine learning models. For various reasons that will be explained here, we propose to use the {tidymodels}[8] framework for our analysis.

Similar to the {tidyverse}, {tidymodels} is a collection of packages dedicated to modeling. It contains packages such as {rsample} (general resampling infrastructure), {yardstick} (performance metrics), {recipes} (preprocessing and feature engineering steps for modeling), {workflows} (modeling workflow), {broom} (tidy statistical objects), and {parsnip} (fitting models) just to name a few. Yet, the similarity between {tidymodels} and {tidyverse} does not end there since {tidymodels} is built (and uses) on the {tidyverse}, hence being the perfect extension for modeling data.

Besides modeling data, {tidymodels} aims at *tidying* the process of modeling data. Such process is done at different levels:

- Tidying the entire modeling workflow by integrating the different steps (including data preparation, model fitting, and data prediction) into simple functions ({parnsip}).
- Tidying (by standardizing) the inputs and outputs for the different Machine Learning algorithms.[9]
- Tidying the models so that the outputs can be easily extracted and used.
- Providing all the relevant functions required for modeling in one unique collection of packages.

Regardless of the algorithm used, the typical modeling approach used by {tidymodels} is as follows:

1. Split your data into training and test set (including sets for Cross-Validation);

[8] https://www.tidymodels.org/

[9] To avoid reinventing the wheel and to be more flexible, {tidymodels} allows calling ML algorithm from various packages in a standardized way, even when those packages often require the data to be structured in a different way, use different names for similar parameters, etc.

2. Build a recipe by informing the model and any preprocessing step required on the data;
3. Define the model (and its parameter) to consider;
4. Create a workflow by combining the previous step together;
5. Run your model;
6. Evaluate your model;
7. Predict new values.

For more information, we refer the readers to *Tidy Modeling with R*[10] by Max Kuhn and Julia Silge.

Let's load the {`tidymodels`} package:

```
library(tidymodels)
```

### 12.6.2 Sampling the Data

As mentioned earlier, an important step consists of splitting the data into a training and testing set. To do so, the function `initial_split()` is used. This function takes as input the original data and returns the information on how to make the different partitions. In practice, such partition could be obtained completely randomly by simply specifying the proportion of data in each partition (here `prop=0.7` meaning that 70% of the data are in the training set and the rest being in the test set). However, we can provide constraints so that the structure of the original data is respected. In our case, `Type` contains three levels which may not be perfectly balanced. By specifying `strata=Type`, we ensure that the different splits respect the original data in terms of proportions for `Type`.

After the `initial_split()`, the `training()` and `testing()` functions are used to obtain the training and testing subsets.

```
wine_split <- initial_split(data=wine, strata="Type", prop=0.7)
wine_train <- training(wine_split)
wine_testing <- testing(wine_split)
```

### 12.6.3 Cross-Validation

Cross-validation (CV) is an important step for checking the model quality. To allow performing CV, additional sets of data are required. These sets of data

[10] https://www.tmwr.org/

can be obtained through the resampling method before building the model. In practice, for each new set of data, a subset is used for building the model, the other subset being then used to measure the performance of such model (similar to the training and testing set defined earlier). However, in this case, the resampling is only performed on the training set defined earlier.

To generate such sets of data, the `vfold_cv` function is used. Here we start with a five-fold cross-validation first. Again, `strata=Type` to a conduct stratified sampling to ensure that each resample is created within the stratification variable.

```
wine_cv <- wine_train %>%
  vfold_cv(v=5, strata=Type)
```

### 12.6.4 Data Preprocessing {recipes}

The {recipes} package contains a rich set of data manipulation tools which can be used to preprocess the data and to define roles for each variable (e.g. outcome and predictor). To add a recipe, the function `recipe()` is used. This function has two arguments: a formula and the data (here `wine_train`). Any variable on the left-hand side of the tilde (~) is considered the model outcome. In our example, we want to use a machine learning model to predict the type of the wine; therefore, `Type` would be the target on the left-hand side of the ~. On the right-hand side of the tilde are the predictors. One can write out all the variables, but an easier option is to use the dot (`.`) to indicate all other variables as predictors.

```
model_recipe <- wine_train %>%
  recipe(Type ~ .)
```

In this instance, we do not need to preprocess further any of the variables present in the data. Yet, if it was the case, we could use the various `step_*()` functions, which then perform any transformation required on the declared variables including:

- `step_log()` for a log transformation;
- `step_dummy()` to transform categorical data into dummy variables (useful to combine with functions such `all_nominal_predictors()`, `starts_with()`, `matches()`, etc.);
- `step_interact()` creates interaction variables;
- `step_num2factor()` converts numeric variables to factor;

- `step_scale()` scales numeric variables;
- `step_pca()` converts numeric data into 1 or more principal components, etc.

---

It should be noted that, by default, any of the preprocessing performed here on the trained data set is also applied adequately on the test set. For instance, with `step_scale()`, the mean and standard deviation are computed on the training data and are then applied on the test data (the means and standard deviation are not recomputed from the test set).

---

### 12.6.5 Model Definition

Once the model formula is defined and the instructions for the data preprocessing is set, we need to decide which type of ML algorithm should be used. Let's consider the random forest classifier for the wine data using `rand_forest()` (the algorithm proposed by the {ranger} package is used here). This function has three hyper-parameters (`mtry, trees, and min_n`) which can be tuned to achieve the best possible results.

Model tuning is the process of finding the optimal values for those parameters. In order to find the best hyper-parameter combinations, we need to define a search range for each of them. When we choose the family of the model we want to use (`rand_forest` in this example), we have to let the machine know that a given parameter (`mtry, trees, min_n`) is not defined explicitly and will be tuned instead. To achieve such a result, we must use the function `tune()`.

```
rf_spec <- rand_forest(
  mtry = tune(),
  trees = tune(),
  min_n = tune()) %>%
  set_mode("classification") %>%
  set_engine(engine = "ranger")
```

### 12.6.6 Set the Whole Process into a Workflow

Finally, we combine the model and the recipe into a single `workflow()`:

```
rf_wf <- workflow() %>%
  add_recipe(model_recipe) %>%
  add_model(rf_spec)
```

### 12.6.7 Tuning the Parameters

In the previous section, placeholders for tuning hyper-parameters were created. It is time to define the scope of the search and to choose the method for searching the parameter space. To do so, `grid_regular()` can be used:

```
params_grid <- rf_spec %>%
  parameters() %>%
  update(mtry = mtry(range = c(1,2)),
         trees = trees(range = c(10,200))) %>%
  grid_regular(levels=5)
```

Now that the hyper-parameter search range is defined, let's look for the best combination using the `tune_grid()` function. The cross-validation set is used for this purpose, so that the data used for training the model has not been used yet.

```
tuning <- tune_grid(rf_wf, resamples=wine_cv, grid=params_grid)
```

The `autoplot()` function is called to take a quick look at the `tuning` object (Figure 12.4).

```
autoplot(tuning)
```

Ultimately, the best combination of parameters is obtained using the `select_best()` function. Such paramaters are defined based on the quality of the model, which can be estimated through a various metrics. Here we decided to use `roc_auc` (Area Under the Receiver Operating Characteristic Curve), as it provides a reliable estimate of the quality of the model.

```
params_best <- select_best(tuning, "roc_auc")
```

### 12.6.8 Model Training

The best parameters can be applied to our model, and the final model can be trained using the entire training set. This is done using the `fit()` function that we apply to our workflow.

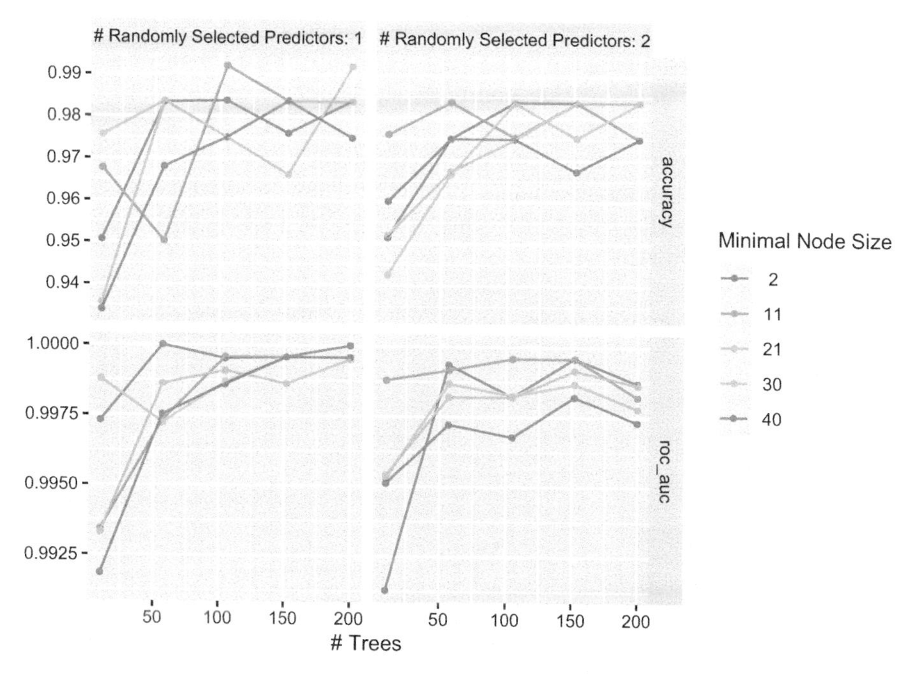


**FIGURE 12.4**
Tuning plot.

```
final_model <- rf_wf %>%
  finalize_workflow(params_best) %>%
  fit(wine_train)
```

### 12.6.9 Model Evaluation

A very important part in building machine learning models is to assess the quality of the model. A first approach consists of applying the model thus obtained on the testing data set (`wine_testing`), which the model has not seen yet.

To do so, the `predict()` function is used. The `predict()` function of {`tidymodels`} allows adding in an easy way the predictions obtained from models to the original data. This procedure allows comparing the predictions with the actual data:

```
obs_vs_pred <- wine_testing %>%
  bind_cols(predict(final_model, .))
```

Here, `obs_vs_pred` is a data frame which contains both the actual wine type (`Type`) and the predicted wine type (`.pred_class`). Comparing these two variables allow judging the quality of the model. Such comparison can be done through a confusion matrix (Figure 12.5). A confusion matrix is a table where each row represents instances in the actual class, while each column represents the instances in a predicted class. From the `autoplot()` function, it appears that the predictions were almost perfect (only two wines were wrongly classified).

```
cm <- conf_mat(obs_vs_pred, Type, .pred_class)

autoplot(cm, type = "heatmap")
```

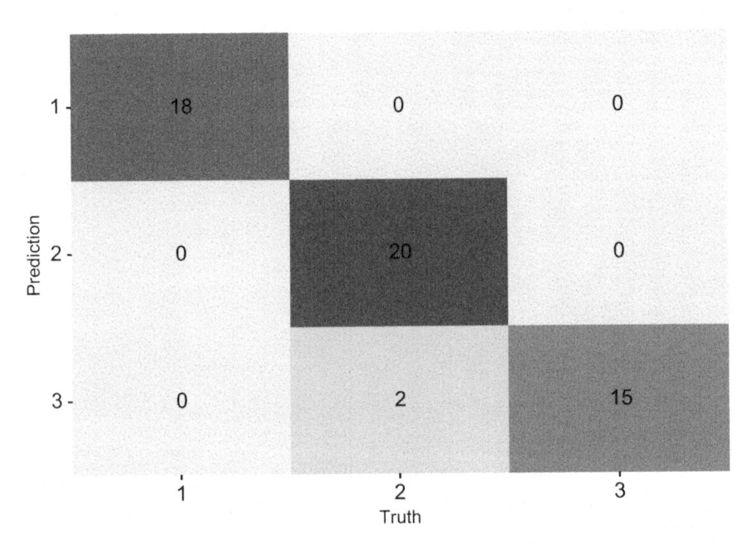


**FIGURE 12.5**
Confusion matrix.

There are several ways to look into the model quality, with the approach and model metrics highly dependent on the situation/type of model used. In our case, as we have a multiclass classification, we can besides use the confusion matrix, to check the model accuracy. Classification accuracy is a metric that summarized the fraction of predictions our model got right (somehow it brings similar information we get from the confusion matrix). We have to first organize the data to have the model predictions. Then we can directly calculate the accuracy and also kappa accuracy, using the functions `accuracy ()` and `kap ()`. Kappa (kap()) is a similar measure to accuracy(), but is normalized by the accuracy that would be expected by chance alone. It can be very useful when one or more classes have large frequency distributions.

```
obs_vs_pred_prob <- bind_cols(wine_testing %>%
                                select(Type),
                              predict(final_model,
                                      wine_testing,
                                      type = "prob")) %>%
  mutate(Type = as.factor(Type))

accuracy(obs_vs_pred, truth = "Type", estimate = ".pred_class")
```

```
## # A tibble: 1 x 3
##   .metric  .estimator .estimate
##   <chr>    <chr>          <dbl>
## 1 accuracy multiclass     0.964
```

```
kap(obs_vs_pred, truth = "Type", estimate = ".pred_class")
```

```
## # A tibble: 1 x 3
##   .metric .estimator .estimate
##   <chr>   <chr>          <dbl>
## 1 kap     multiclass     0.945
```

The accuracy and Kappa accuracy of our model is extremely high ($> 0.94$), emphasizing its great model performance.

A simple and easy way to have a higher level understanding on what variables play the most important role in our model (wine classification) is through the Feature Importance plot. To create this plot, we need to first create an object known as `explainer` from the tidymodels workflow, using the function `explain_tidymodels` from the package {`DALEXtra`}. This function will take as argument: the model to be explained, the data to be used to calculate the explanations which should be passed without a target column (`wine_train` removing the columns `Type`) and the numeric vector with outputs/scores (y). The `explainer` object can be then used to create the Feature Importance plot (Figure 12.6) using the function `variable_importance`.

The interpretation is very straightforward, in the way that variables are conveniently ordered according to their importance. The higher the cross-entropy loss after permutations, the more important the variable is to, in our case, decide to which group each wine belong to. So, for example, if color is permuted (spoil the variable), it turns out that the model will be more than 2.5 times worse than the one with the correct color variable. In summary, cross-entropy loss after permutations measures how much the permutation of a variable would impact the model performance. The higher the impact, the most important the variable.

```
library(DALEXtra)
library(modelStudio)

data_to_explain <- wine_train

explainer <- explain_tidymodels(model = final_model,
                                data = data_to_explain %>%
                                  select(-Type),
                                y = data_to_explain$Type)

var_imp <- variable_importance(explainer,
                               loss_function = loss_cross_entropy,
                               type = "ratio")
plot(var_imp)
```

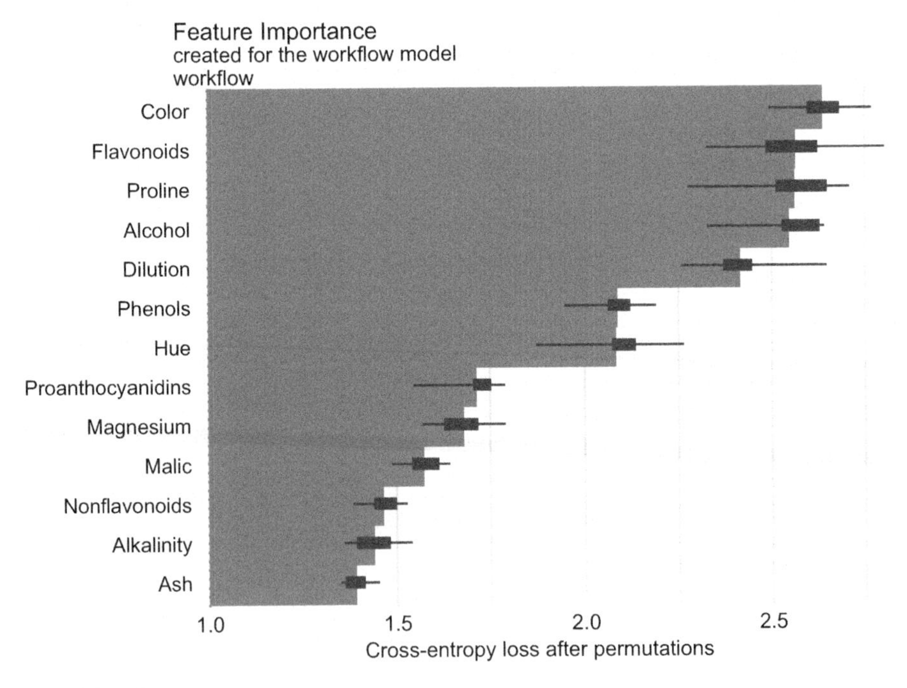


**FIGURE 12.6**
Feature importance.

Building a machine learning model may seem complicated at first since there are many steps and important things to watch out for. Although we presented here a simple example, for one specific situation of classification, you will realize that this way of working (typical modeling approach showed) is highly applicable to other situations. Again, the main idea of this chapter is to open your mind, provide you with the basics, and ultimately motivate you to seek and learn more about machine learning!

# 13

# *Text Analysis*

In the previous chapters, most transformations and analyses were performed on *simple* data, that is, data that represent something very specific, understandable, predictable, and stand-alone. For numerical variables (e.g. sensory attributes), one data entry is simply a number often defined within a range. For categorical variables or factors, each data entry is a predefined entry (e.g. product names or a category for a given variable) chosen from a list of possible options. But there are situations where the data are intrinsically more *complex* and less structured. A good illustration of such *complex* situation is text analysis. Before collecting the data, we do not know explicitly what kind of information we will get (with open-ended questions, respondents are free to say/write whatever they want!). In that case, each data entry (from words, to sentences, to paragraphs...) is more *messy* as it may contain relevant and less-informative elements. The goal of the analysis is then to extract the relevant information from the data and to summarize it automatically. In this section, we will show you how such data can be processed and how information can be extracted.

## 13.1 Introduction to Natural Language Processing

Humans exchange information through the use of languages. There is of course a very large number of different languages, each of them having their own specificity. The science that studies languages per se is called *linguistics*: It focuses on areas such as phonetics, phonology, morphology, syntax, semantics, and pragmatics.

Natural Language Processing (NLP) is a subfield of linguistics, computer science, and artificial intelligence. It connects computers to human language by processing, analyzing, and modeling large amounts of natural language data. One of the main goals of NLP is to *understand* the contents of documents

DOI: 10.1201/9781003028611-13

and to extract accurately information and insights from those documents. In Sensory and Consumer Research, we often refer to NLP when we talk about *Text Analysis*.

Since the fields of linguistics and NLP are widely studied, a lot of documentations are already available online. The objective of this chapter is to provide sufficient information for you to be familiar with textual data and to give you the keys to run the most useful analyses in Sensory and Consumer Research.

For those who would like to dive deeper into NLP, we recommend reading Silge and Robinson (2017), Bécue-Bertaut (2019), and Hvitfeldt and Silge (2021) for more advanced techniques.

## 13.2 Application of Text Analysis in Sensory and Consumer Science

### 13.2.1 Text Analysis as Way to Describe Products

In recent years, open-ended comments have gained interest as it is the fastest, safest, and most unbiased way to collect spontaneous data from participants (Piqueras-Fiszman, 2015).

Traditionally, most SCS questionnaires relied primarily on closed questions, to which open-ended questions were added to uncover the consumers' reasons for liking or disliking products. In practice, these open-ended questions were positioned right after liking questions and aimed at providing some understanding about why a product may or may not be liked and to give the participants a chance to reduce their frustration by explaining their responses to certain questions. As a result of such practices, these questions were usually not deeply analyzed.

With the development of the so-called *rapid* and consumer-oriented descriptive methods, the benefits of open-ended questions became more apparent as they provide a new way to uncover sensory perception. In practice, respondents are asked to give any terms that describe their sensory perception in addition to their quantitative evaluation of the products by means of intensity rating or ranking (e.g. Free Choice Profile, Williams and Langron, 1984; Flash Profile, Delarue and Sieffermann, 2004), or similarities and dissimilarities assessment (e.g. Free Sorting Task, Cadoret et al., 2009; Ultra Flash Profile, Perrin et al., 2008). Since the textual responses are now an integral part of the method, their analysis can no longer be ignored.

The importance of open-ended questions increased further as it has been shown that respondents can reliably describe in their own words their full experience (perception, emotion, or any other sort of association) with products. Recently, Mahieu et al. (Mahieu et al., 2020, 2022; Visalli et al., 2020) showed the benefits of using open-ended questions over CATA.[1] In this study, consumers were asked to describe with their own words both the products they evaluated and what their ideal product would be like. Similarly, Luc et al. (2020, 2022a,b) proposed an alternative to the Just About Right (JAR) scale method – called free-JAR – and in which consumers describe the samples using their own words, by still following a JAR terminology (too little, too much, or JAR, etc.).

The inclusion of open-ended questions as one of the primary elements of sensory and consumer tasks blurs the line with other fields, including psychology and sociology where these qualitative methods originated. More recently, advances in the technology (web-scraping, social listening, etc.) opened new doors that brought SCS closer to other fields such as marketing for instance. The amount of data that are collected with such techniques can be considerably larger, but the aim of the analysis stays the same: extracting information from text/comments. Various examples of the application of text analysis in the SCS field can be found in Bécue-Bertaut et al. (2008), ten Kleij and Musters (2003), Hamilton and Lahne (2020), Vidal et al. (2015), or Feldmeyer and Johnson (2022) just to name a few.

### 13.2.2 Objectives of Text Analysis

Open-ended comments, and more generally textual responses in questionnaires, are by definition qualitative. This means that the primary analysis should be qualitative. It could simply consist of reading all these comments and eventually summarizing the information gathered. But as the number of comments increases, such an approach quickly becomes too time and energy consuming for the analysts. How can we transform such qualitative data into quantitative measures? How can we digest and summarize the information contained in these comments without losing the overall meaning of the messages (context)?

One easy solution is to simply count how often a certain word is being used in a given context (e.g. how often the word `sweet` is being associated to each product evaluated). However, if such a solution is a reasonable one to start with, we will show some alternatives that allow going deeper into the understanding of textual inputs. This is the objective of the textual analysis and NLP that we are going to tackle in the next sections.

[1] CATA can be seen as a simplified version of open-comments in the sense that respondents also associate products to words; however, they lose the freedom of using their own as they need to select them from a predefined list.

### 13.2.3 Classical Text Analysis Workflow

In SCS, the generic notion of *text analysis* often includes any step or procedure that allows going from the raw data (e.g. consumer comments, text scrapped from website or social media, etc.) to results and insights. However, such process requires many separate steps, often defined as follows:

1. **Tokenization** is the step that splits the raw data into statistical units of interest, also called token.[2]
2. Non-informal words or **stopwords** (e.g. *and*, *I*, *you*, etc.) are then removed from the data to facilitate the extraction of the information.
3. **Stemming** consists of reducing words to their root form, hence grouping the different variants of the same word (e.g. singular/plural, infinitive or conjugated verbs, etc.).
4. An extra (optional) step called **lemmatization** consists of grouping words that have similar meanings under one umbrella. The advantage of such procedure is that it simplifies further the analysis and its interpretation. However, it can be time consuming and more importantly, it relies on the analyst's own judgment: two different analysts performing the same task on the same data will obtain different end results.
5. The final data are then **analyzed** and summarized (often through counts) to extract information or patterns.

### 13.2.4 Warnings

Languages are complex, as many aspects can influence the meaning of a message. For instance, in spoken languages, the intonation is as important as the message itself. In written languages, non-word items (e.g. punctuation, emojis) may also completely change the meaning of a sentence (e.g. irony). Worst, some words have different meanings depending on their use (e.g. *like*), and the context of the message provides its meaning. Unfortunately, the full *context* is only available when analyzed manually (e.g. when the analyst reads all the comments), meaning that automating analyses do not always allow capturing it properly. In practice, however, reading all the comments is not a realistic solution. This is why we suggest to automate the analysis to extract as much information as possible, before going back to the raw text to ensure that the conclusions drawn match the data.

---

[2] A token can be a single word, a group of $n$-words (also know as *n-grams*), a sentence, or an entire document.

## 13.3 Illustration Involving Sorting Task Data

Let's start with loading the usual packages of need:

```
library(tidyverse)
library(here)
library(readxl)
```

The data set used for illustration was kindly shared by Dr. Jacob Lahne. It is part of a study that aimed at developing a CATA lexicon for Virginia Hard (Alcoholic) Ciders (Phetxumphou et al., 2020). The data can be found in *cider_text_data.xlsx*.

Let's also import the data to our R session:

```
file_path <- here("data","cider_text_data.xlsx")
cider_og <- read_xlsx(file_path) %>%
  mutate(sample = as.character(sample))
```

### 13.3.1 Data Preprocessing

Before starting, it is important to mention that there is a large variety of R-based solutions and R packages that handle textual data, including:

- The IRaMuTeQ project (CAD1983821870) is a free software dedicated to text analysis and developed in R and Python. It includes Reinert textual clustering method (for more information, see http://www.iramuteq.org/);
- {`tm`} package for text mining;
- {`tokenizers`} to transform strings into tokens;
- {`SnowballC`} for text stemming;
- {`SpacyR`} for Natural Language Processing;
- {`Xplortext`} for deep understanding and analysis of textual data.

However, to ensure a continuity with the rest of the book, we will emphasize the use of the {`stringr`} package for handling strings (here text) combined with the {`tidytext`} package. Note that {`stringr`} is part of the {`tidyverse`} and both packages fit very well within the {`tidyverse`} philosophy.

Let's load this additional package:

```
library(tidytext)
```

### 13.3.2 Introduction to Working with Strings ({stringr})

The {stringr} package brings a large set of tools that allow working with strings. Most functions included in {stringr} start with `str_*()`. Here are some of the most convenient functions:

- `str_length()` to get the length of the string;
- `str_c()` to combine multiple strings into one;
- `str_detect()` to search for a pattern in a string and `str_which()`, find the position of a pattern within the string;
- `str_extract()` and `str_extract_all()` to extract the first (or all) matching pattern from a string;
- `str_remove()` and `str_remove_all()` to remove the first (or all) matching pattern from a string;
- `str_replace()` and `str_replace_all()` to replace the first (or all) matching pattern with another one.

It also includes *formatting* options that can be applied to strings, including:

- `str_to_upper()` and `str_to_lower()` to convert strings to uppercase or lowercase;
- `str_trim()` and `str_squish()` to remove white spaces;
- `str_order` to order the element of a character vector.

Examples of the applications of some of these functions are shown in the next sections.

### 13.3.3 Tokenization

The analysis of textual data starts with defining the statistical unit of interest, also known as *token*. This can either be a single word, a group of words, a sentence, a paragraph, a whole document, etc. The procedure to transform the document into tokens is called *tokenization*.

By looking at our data (`cider_og`), we can notice that for each sample evaluated, respondents are providing a set of responses, ranging from a single word (e.g. `yeasty`) to a group of words (`like it will taste dry and acidic`).

Fortunately, the data are also well structured since the responses seem to be separated by a `;` or `,`.

Let's transform this text into tokens using `unnest_tokens()` from the `{tidytext}` package. The function `unnest_tokens()` proposes different options for the tokenization including *words*, *ngrams*, or *sentences*, for instance. However, let's take advantage of the data structure and use a specific character to separate the tokens (here `;`, `,` etc.). The `regex` parameter allows us to specify the patterns to consider:

```
cider <- cider_og %>%
  unnest_tokens(tokens, comments, token="regex",
                pattern="[;|,|:|.|/]", to_lower=FALSE)
```

The original comments from consumers are now split into tokens, thus increasing the size of the file from 168 individual comments to 947 rows of tokens.

This procedure already provides some interesting information as we could easily count word usage and answer questions such as "how often the word *apple* is used to describe each samples?" for instance. However, a deeper look at the data shows some inconsistencies since some words starts with a space or have capital letters (remember that R is case-sensitive!). Further preprocessing is thus needed.

### 13.3.4 Simple Transformations

To further prepare the data, let's standardize the text by removing all the white spaces (*irrelevant* spaces in the text, e.g. at the start/end, double spaces, etc.), transforming everything to lower case (note that this could have been done earlier through the parameter `to_lower=TRUE` from `unnest_tokens()`), removing some special letters, replacing some misplaced characters, etc.[3]

```
cider <- cider %>%
  mutate(tokens = str_to_lower(tokens)) %>%
  mutate(tokens = str_trim(tokens)) %>%
  mutate(tokens = str_squish(tokens)) %>%
  mutate(tokens = str_remove_all(tokens, pattern="[(|)|?|!]")) %>%
  mutate(tokens = str_remove_all(tokens, pattern="[ó|ò]")) %>%
  mutate(tokens = str_replace_all(tokens, pattern="õ", replacement="'"))
```

[3] This process is done in iterations: the more you clean your document, the more you find some small things to fix... until you're set!

To ensure that the cleaning job is done (for now), let's produce the list of tokens generated here (and its corresponding frequency)[4]:

```
cider %>%
  count(tokens) %>%
  arrange(desc(n))
```

```
## # A tibble: 476 x 2
##   tokens      n
##   <chr>   <int>
## 1 sweet      55
## 2 fruity     33
## 3 sour       32
## 4 tart       28
## # ... with 472 more rows
```

The most used words to describe the ciders are `sweet` (55 occurrences), `fruity` (33 occurrences), and `sour` (32 occurrences).

A closer look at this list highlights a few things that still need to get tackled:

- The same concept can be described in different ways: `spicy`, `spices`, and `spiced` may all refer to the same concept, yet they are written differently and hence are considered as different tokens. This will be handled in a later stage.
- Multiple concepts are still joined and hence considered separately: `sour and sweet` is currently neither associated to `sour` nor to `sweet`, and we may want to disentangle them.
- There could be some typos: Is `sweat` a typo and should read `sweet`? Or did that respondent really perceived the cider as `sweat`?
- Although most tokens are made of one (or few) words, some others are defined as a whole sentence (e.g. `this has a very lovely floral and fruity smell`).

Let's handle each of these different points...

### 13.3.5 Splitting Further the Tokens

For an even deeper cleaning, let's go one step further and split the remaining tokens into single words by using the space as separator. Then, we can number each token for each assessor using `row_number()` to ensure that we can still

[4] Although not present in the text, we will use the next three lines of code multiple times to count the number of words present in the data.

recover which words belong to the same token, as defined previously. This information will be specially relevant later when looking at *bigrams*.

```
cider <- cider %>%
  relocate(subject, .before=sample) %>%
  group_by(subject, sample) %>%
  mutate(num = row_number()) %>%
  ungroup() %>%
  unnest_tokens(tokens, tokens, token="regex", pattern=" |-")

head(cider)
```

```
## # A tibble: 6 x 5
##   subject sample rating tokens      num
##   <chr>   <chr>   <dbl> <chr>     <int>
## 1 J1      182         8 hard          1
## 2 J1      182         8 cider         1
## 3 J1      182         8 smell         1
## 4 J1      182         8 fermented     2
## # ... with 2 more rows
```

For `J1` and `182`, for instance, the first token is now separated into three words: `hard`, `cider`, and `smell`.

A quick count of words show that `sweet` appears now 96 times and `apple` 82 times. Interestingly, terms such as `a`, `like`, `the`, `of`, `and`, etc. also appear fairly frequently.

### 13.3.6 Stopwords

*Stop words* refer to *common* words that do not carry much (if at all) information. In general, stop words include words (in English) such as *I*, *you*, *or*, *of*, *and*, *is*, *has*, etc. It is thus common practice to remove such stop words before any analysis as they would *pollute* the results with unnecessary information.

Building lists of stop words can be tedious. Fortunately, it is possible to find some predefined lists and to eventually adjust them to our own needs by adding and/or removing words. In particular, the package {`stopwords`} contains a comprehensive collection of stop word lists:

```
library(stopwords)
length(stopwords(source="snowball"))
```

```
## [1] 175
```

```
length(stopwords(source="stopwords-iso"))
```

```
## [1] 1298
```

The English Snowball list contains 175 words, whereas the English list from the Stopwords ISO collection contains 1298 words.

A deeper look at these lists (and particularly to the *Stopwords ISO* list) shows that certain words including *like*, *not* and *don't* (just to name a few) are considered as stop words. If we would use this list blindly, we would remove these words from our comments. Although using such list on our current example would have a limited impact on the analysis (most comments are just few descriptive words), it would have a more critical impact on other studies in which consumers give their opinion on samples. Indeed, the analysis of the two following comments *I like Sample A* and *I don't like Sample B* would be lost although they provide some relevant information.

It is therefore important to remember that although a lot of stop words are relevant in all cases, some of them are topic specific and should (or should not) be used in certain contexts. Hence, inspecting and adapting these lists before use is strongly recommended.

Since we have a relatively small text size, let's use the *SnowBall Stopword* list as a start, and look at the terms that our list and this stopword list share:

```
stopword_list <- stopwords(source="snowball")
word_list <- cider %>%
  count(tokens) %>%
  pull(tokens)

intersect(stopword_list, word_list)
```

As we can see, some words such as *off*, *not*, *no*, *too*, and *very* would automatically be removed. However, such qualifiers are useful in the interpretation of sensory perception, so we would prefer to keep them. We can thus remove them from `stopword_list`.

```
stopword_list <- stopword_list[!stopword_list %in%
                                 c("off","no","not","too","very")]
```

Conversely, we can look at the words from our data that we would not consider relevant and add them to the list. To do so, let's look at the list of words in our data that is not present in `stopword_list`:

```
word_list[!word_list %in% stopword_list]
```

Words such as *like*, *sample*, *just*, *think*, or *though* do not seem to bring any relevant information here. Hence, let's add them (together with others) to our customized list of stop words[5]:

```
stopword_list <- c(stopword_list,
                   c("accompany","amount","anything","considering",
                     "despite","expected","just","like","neither",
                     "one","order","others","products","sample",
                     "seems","something","thank","think","though",
                     "time","way"))
```

A final look at the list of stop words (here ordered alphabetically) ensures that it fits our need:

```
stopword_list[order(stopword_list)]
```

Finally, the data are being cleaned by removing all the words stored in `stopword_list`. This can easily be done either using `filter()` (we keep tokens that are not contained in `stopword_list`) or by using `anti_join()`[6]:

```
cider <- cider %>%
  anti_join(tibble(tokens = stopword_list), by="tokens")
```

### 13.3.7 Stemming and Lemmatization

After removing the stop words, the data contain a total of 328 different words. However, a closer look at this list shows that it is still not optimal, as for instance `apple` (82 occurrences) and `apples` (24 occurrences) are considered as two separate words although they refer to the same *concept*.

[5] As an exercise, you could go deeper into the list and decide by yourself whether you would want to remove more words.

[6] Note that if we were using the original list of stopwords, `anti_join()` can directly be associated to `get_stopwords(source="snowball")`.

To further *clean* the data, two similar approaches can be considered: **stemming** and **lemmatization**.

The procedure of stemming consists of performing a step-by-step algorithm that reduces each word to its base word (or *stem*). The most used algorithm is the one introduced by Porter (1980) which is available in the {`SnowballC`} package through the `wordStem()` function:

```
library(SnowballC)

cider <- cider %>%
  mutate(stem = wordStem(tokens))
```

The stemming reduced further the list to 303 words. Now, `apple` and `apples` have been combined into `appl` (106 occurrences). However, due to the way the algorithm works, the final tokens are no longer English[7] words.

Alternatively, we can *lemmatize* words. Lemmatization is similar to stemming except that it does not cut words to their stems: Instead it uses knowledge about the language's structure to reduce words down to their dictionary form (also called *lemma*). This approach is implemented in the {`spacyr`} package[8] and the `spacy_parse()` function:

```
library(spacyr)

spacy_initialize(entity=FALSE)
lemma <- spacy_parse(cider$tokens) %>%
  as_tibble() %>%
  dplyr::select(tokens=token, lemma) %>%
  unique()

cider <- full_join(cider, lemma, by="tokens")
```

As can be seen, as opposed to stems, lemmas consist of *regular* words. Here, the grouping provides similar number of terms (approx 300) in both cases:

[7] Different algorithms for different languages exist, so we are not limited to stemming English words.

[8] spaCy is a library written in Python: for the {`spacyr`} package to work, you'll need to go through a series of steps that are described here: (https://cran.r-project.org/web/packages/spacyr/readme/README.html)[https://cran.r-project.org/web/packages/spacyr/readme/README.html]

```
cider %>% count(stem)
```

```
## # A tibble: 301 x 2
##   stem          n
##   <chr>     <int>
## 1 acid         23
## 2 acrid         1
## 3 aftertast    12
## 4 alcohol      13
## # ... with 297 more rows
```

```
cider %>% count(lemma)
```

```
## # A tibble: 303 x 2
##   lemma       n
##   <chr>   <int>
## 1 acid        3
## 2 acidic     18
## 3 acidity     2
## 4 acrid       1
## # ... with 299 more rows
```

In the case of lemmatization, `acid`, `acidity`, and `acidic` are still considered as separate words whereas they are all grouped under `acid` with the stemming procedure. This particular example shows the advantage and disadvantage of each method, as it may (or may not) group words that are (or are not) meant to be grouped. Hence, the use of lemmatization/stemming procedures should be thought carefully. Depending on their objective, researchers may be interested in the different meanings conveyed by such words as `acid`, `acidity`, and `acidic` and decide to keep them separated or decide to group them for a more holistic view of the main sensory attributes that could be derived from this text.

It should also be said that neither the lemmatization nor the stemming procedure will combine words that are different but bear similar meanings. For instance, the words `moldy` and `rotten` have been used, and some researchers may decide to group them if they consider them equivalent. This type of grouping should be done manually on a case-by-case using `str_replace()`:

```
cider %>%
  count(lemma) %>%
  filter(lemma %in% c("moldy","rotten"))
```

```
## # A tibble: 2 x 2
##   lemma      n
##   <chr>  <int>
## 1 moldy      2
## 2 rotten     5
```

As can be seen here, originally, `moldy` was stated twice whereas `rotten` was stated five times. After replacing `moldy` by `rotten`, the newer version contains seven occurrences of `rotten` and none of `moldy`.

```
cider %>%
  mutate(lemma = str_replace(lemma, "moldy", "rotten")) %>%
  count(lemma) %>%
  filter(lemma %in% c("moldy","rotten"))
```

```
## # A tibble: 1 x 2
##   lemma      n
##   <chr>  <int>
## 1 rotten     7
```

Doing such transformation can quickly be tedious to do directly in R. As an alternative solution, we propose to export the list of words in Excel, create a new column with the new grouping names, and merge the newly acquired names to the previous file. This is the approach we used to create the file entitled *Example of word grouping.xlsx*. In this example, one can notice that we limited the grouping to a strict minimum for most words except `bubble` that we also combined to `bubbly`, `carbonate`, `champagne`, `moscato`, `fizzy`, and `sparkle`:

```
new_list <- read_xlsx("data/Example of word grouping.xlsx")
cider <- cider %>%
  full_join(new_list, by="lemma") %>%
  mutate(lemma = ifelse(is.na(`new name`), lemma, `new name`)) %>%
  dplyr::select(-`new name`)
```

This last *cleaning* approach reduces further the number of words to 278.

## 13.4 Text Analysis

Now that the text has been sufficiently cleaned, some analyses can be run to compare the samples in the way they have been described by the respondents. To do so, let's start with simple analyses.

### 13.4.1 Raw Frequencies and Visualization

In the previous sections, we have already shown how to count the number of occurrences of each word. We can reproduce this and show the top 10 most used words to describe our ciders:

```
cider %>%
  group_by(lemma) %>%
  count() %>%
  arrange(desc(n)) %>%
  filter(n>=10, !is.na(lemma)) %>%
  ggplot(aes(x=reorder(lemma, n), y=n))+
  geom_col()+
  theme_minimal()+
  xlab("")+
  ylab("")+
  theme(axis.line = element_line(colour="grey80"))+
  coord_flip()+
  ggtitle("List of words mentioned at least 10 times")
```

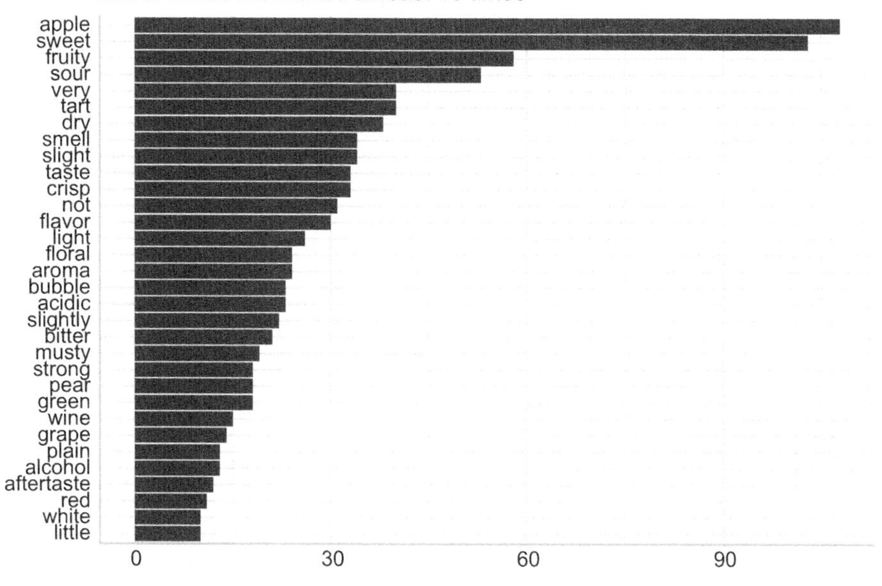


As seen previously, the most mentioned words are `apple`, `sweet`, `fruity`, and `sour`.

Let's now assess the number of time each word has been used to characterize each product.

```
cider %>%
  filter(!is.na(lemma), !is.na(sample)) %>%
  group_by(sample, lemma) %>%
  count() %>%
  ungroup() %>%
  pivot_wider(names_from=lemma, values_from=n, values_fill=0)
```

```
## # A tibble: 6 x 276
##   sample acidic aftertaste alcohol appeal apple aroma
##   <chr>   <int>      <int>   <int>  <int> <int> <int>
## 1 182         6          2       3      1    18     4
## 2 239         5          2       2      2    19     5
## 3 365         3          3       0      0    25     3
## 4 401         4          2       2      0     9     5
## # ... with 2 more rows, and 269 more variables:
## #   artificial <int>, astringent <int>, bad <int>,
## #   banana <int>, barn <int>, begin <int>,
## #   bitter <int>, blackberry <int>, bland <int>,
## #   bold <int>, bubble <int>, candy <int>,
## #   cider <int>, clean <int>, crisp <int>,
## #   decent <int>, different <int>, dog <int>, ...
```

A first look at the contingency table shows that `apple` has been used 25 times to characterize sample `365` while it has only been used 9 times to characterize sample `401`.

Since the list of terms is quite large, we can visualize these frequencies in different ways: First, we could readapt the histogram produced previously overall but per product. This could give a good overview of which words characterize each sample (results not shown here):

```
prod_term <- cider %>%
  filter(!is.na(lemma), !is.na(sample)) %>%
  group_by(sample, lemma) %>%
  count() %>%
  ungroup() %>%
  split(.$sample) %>%
  map(function(data){
    data %>%
      arrange(desc(n)) %>%
      filter(n>=5) %>%
      ggplot(aes(x=reorder(lemma, n), y=n))+
      geom_col()+
      theme_minimal()+
      xlab("")+
      ylab("")+
      theme(axis.line = element_line(colour="grey80"))+
```

```
        coord_flip()+
        ggtitle(paste0("List of words mentioned at least 5 times for ",
                       data %>% pull(sample) %>% unique()))
    })
```

Another approach consists of visualizing the association between the samples and the words using Correspondence Analysis (CA). Since the CA can be sensitive to low frequencies (see Chapter 2 in Husson et al., 2017), we suggest to only keep terms that were at least mentioned five times across all samples, resulting in a shorter frequency table. We then use the `CA()` function from {`FactoMineR`} to build the CA map:

```
cider_ct <- cider %>%
  filter(!is.na(lemma), !is.na(sample)) %>%
  group_by(sample, lemma) %>%
  count() %>%
  ungroup() %>%
  filter(n >= 5) %>%
  pivot_wider(names_from=lemma, values_from=n, values_fill=0) %>%
  as.data.frame() %>%
  column_to_rownames(var="sample")

library(FactoMineR)
cider_CA <- CA(cider_ct)
```

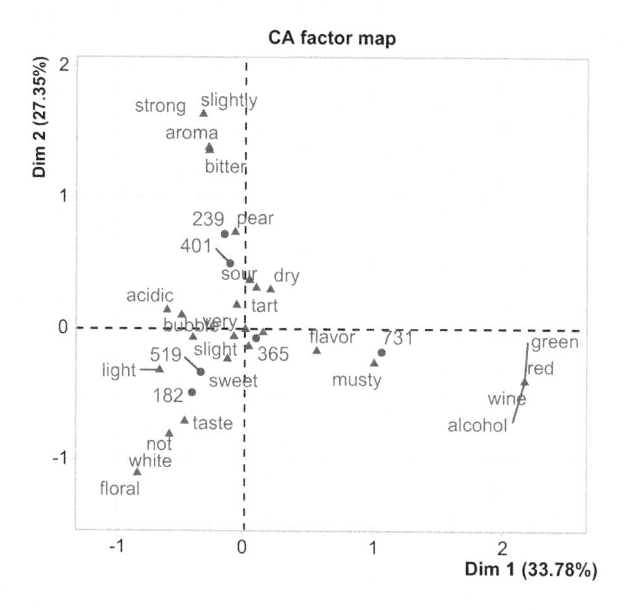

As can be seen, sample `731` is more strongly associated with alcoholic terms such as `alcohol` or `wine`, and colors (`red`, `green`). Samples `239` and `401` are more associated to `sour` and `bitter` (and `pear` for `239`), whereas samples `519` and `182` are more frequently described by terms such as `fruity` and `sweet` (`floral` is also used to characterize `182`).

An alternative for visualizing these frequencies is through wordclouds, which can easily be done using the {`ggwordcloud`} package. This package has the advantage to build such representation in a {`ggplot2`} format. Such wordclouds (here one per product) can be obtained using the following code:

```
cider_wc <- cider %>%
  filter(!is.na(lemma), !is.na(sample)) %>%
  group_by(sample, lemma) %>%
  count() %>%
  ungroup() %>%
  filter(n >= 5)

library(ggwordcloud)
ggplot(cider_wc, aes(x=sample, colour=sample, label=lemma, size=n))+
  geom_text_wordcloud(eccentricity = 2.5)+
  xlab("")+
  theme_minimal()
```

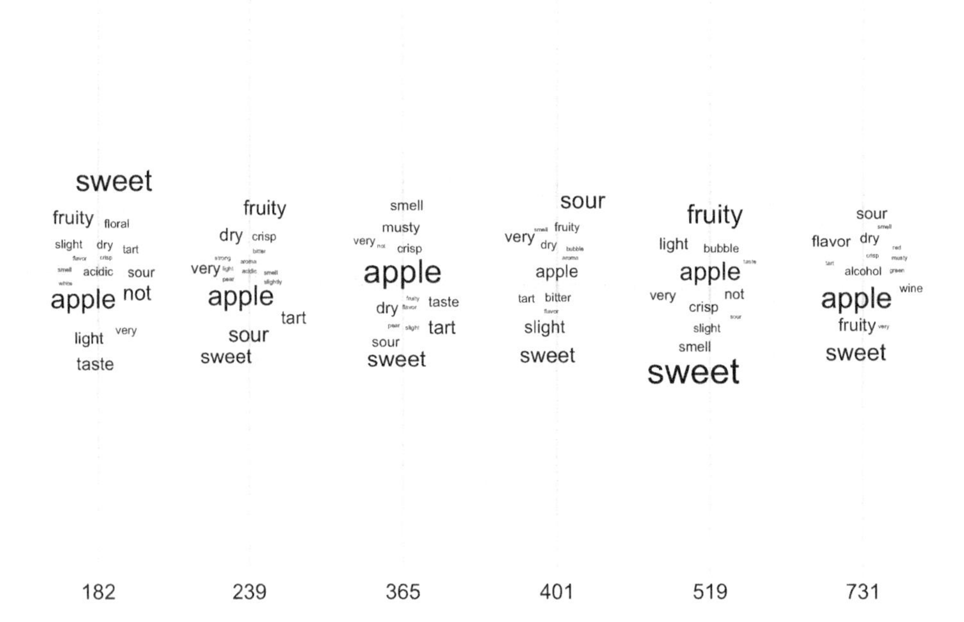

In these wordclouds, we notice that `apple` and `sweet` appear in larger fonts for (almost) all the samples, which can make the comparison quite difficult between samples. Fortunately, the `geom_text_wordcloud()` function provides an interesting parameter in its aesthetics called `angle_group` which allows controlling the position of the words. To illustrate this, let's apply the following rule: for a given sample, if the proportion of association of a word is larger than 1/6 (as we have six samples), the word will be printed in the upper part of its wordcloud, and in the lower part otherwise. To facilitate readability, the color code used followed the same rule:

```
cider_wc %>%
  group_by(lemma) %>%
  mutate(prop = n/sum(n)) %>%
  ungroup() %>%
  ggplot(aes(colour= prop<1/6, label=lemma, size=n,
             angle_group = prop < 1/6))+
  geom_text_wordcloud(eccentricity = 2.5)+
  xlab("")+
  theme_minimal()+
  facet_wrap(~sample)
```

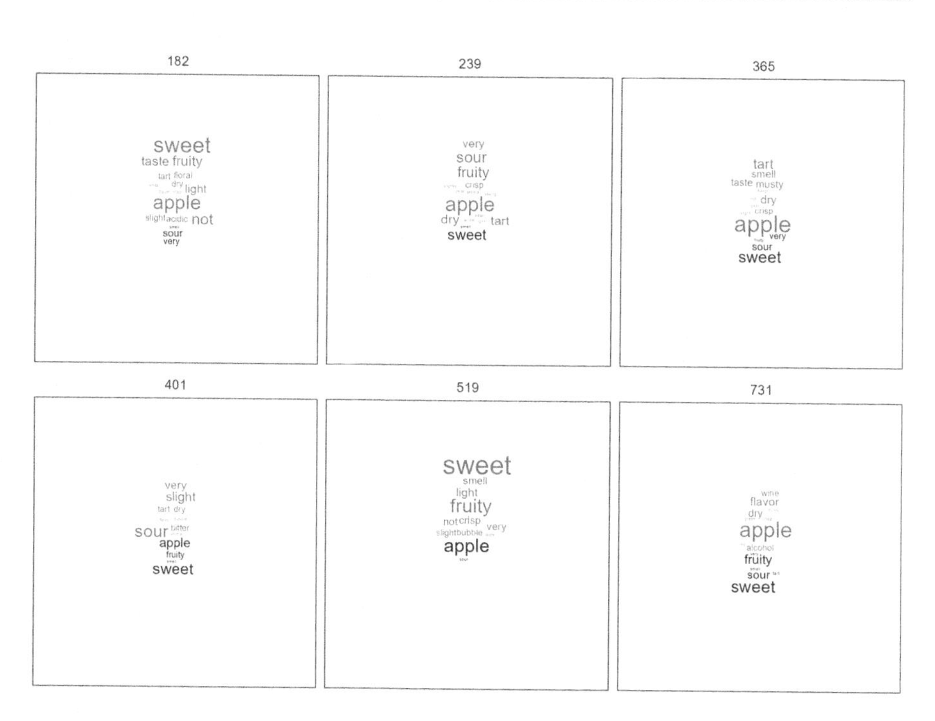

As can be seen, the term apple is more frequently (i.e. more than 1/6) used to characterize samples 182, 239, 365, and 731. The term sweet is more frequently used to characterize samples 182 and 519. Such conclusions would have been more difficult to reach based on the previous *unstructured* wordcloud.

### 13.4.2 Bigrams and n-grams

In the previous set of analyses, we defined each word as a token. This procedure disconnects words from each others, discarding the context around each word. Although this approach is common, it can lead to misinterpretation since a product that would often be associated to (say) *not sweet* would in the end be characterized as *not* and *sweet*. A comparison of samples based on the sole word *sweet* could suggest that the previous product is often characterized as sweet whereas it should be the opposite.

To avoid this misinterpretation, two solutions exist:

1. Replace *not sweet* by *not_sweet*, so that it is considered as one token rather than two;
2. Look at the groups of words, that is, at words within their surroundings.

The latter option leads us to introduce the notion of bi-grams (groups of two following words), tri-grams (groups of three following words), or more generally $n$-grams (groups of $n$ following words). More precisely, we are applying the same frequency count as before except that we are no longer considering one word as a token but as a sequence of 2, 3, or more generally $n$ words as a token. Such grouping can be obtained by the unnest_tokens() from {tidytext} in which token='ngrams', with n defining the number of words to consider.

For simplicity, let's apply this to the original data, although it could be applied to the cleaned version (here we consider bi-grams).

```
cider_2grams <- cider_og %>%
  unnest_tokens(bigrams, comments, token="ngrams", n=2)

cider_2grams %>%
  count(bigrams) %>%
  arrange(desc(n))
```

```
## # A tibble: 1,230 x 2
##   bigrams            n
##   <chr>          <int>
## 1 sweet fruity      11
```

```
## 2 a little        9
## 3 slight apple    9
## 4 smells like     9
## # ... with 1,226 more rows
```

In our example, `sweet fruity` is the strongest two-word association. Other relevant associations are `green apple`, `sweet apple`, or `very sweet`. Of course, such bi-grams can also be obtained per product:

```
cider_2grams %>%
  group_by(sample) %>%
  count(bigrams) %>%
  ungroup() %>%
  arrange(desc(n)) %>%
  filter(sample == "182")
```

```
## # A tibble: 255 x 3
##   sample bigrams          n
##   <chr>  <chr>        <int>
## 1 182    hint of          3
## 2 182    not sweet        3
## 3 182    not very         3
## 4 182    red apples       3
## # ... with 251 more rows
```

For sample `182`, `not sweet` appears three times which can be surprising since it was one of the sample the most associated to `sweet` with 22 occurrences.

### 13.4.3 Word Embedding

The previous section introduces the concept of context, as words are associated to their direct neighbors. Another approach called *word embedding* goes one step further by looking at connections between words within a certain window: for instance, how often are *not* and *sweet* present together within a window of 3, 5, or 7 words? Such an approach is not presented here as it is more relevant for longer text documents.

In the previous sections, we already introduced the notion of *term frequency* (*tf*), which corresponds to the number of times a word is being used in a document. When a collection of documents are analyzed and compared, it is also interesting to look at the *inverse document frequency* (*idf*), which consists of highlighting words that discriminate between documents by reducing the weight of common words and by increasing the weight of words that are specific to certain documents only. In practice, both concepts are associated (by multiplication) to compute a term's *tf-idf*, which measures the frequency of a term adjusted for its rarity in use.

### 13.4.4 Sentiment Analysis

Textual analysis as we presented here is purely descriptive. In other words, the items that we analyzed have no particular valence (i.e. they are neither negative nor positive). When text data are more spontaneous (e.g. social media such as tweets or consumers' responses to open-ended questions), they can be charged with positive or negative connotations. A good way to measure the overall valence of a message is through *Sentiment Analysis*.

To perform *Sentiment Analysis*, we start by deconstructing the message into words (tokenization approach considered previously). Then, in a similar approach to the stop words, we can combine our list of words with a predefined list that defines which words should be considered as positive or negative (the rest being neutral). Ultimately, all the scores associated with each message can be summed, hence providing the overall valence score of a message.

To get examples of *sentiment* list, the `get_sentiments()` function from the `{tidytext}` package can be used. This function proposes four potential lists: `"bing"`, `"afinn"`, `"loughran"`, and `"nrc"` (see Liu, 2015; Mohammad and Turney, 2013 for example). Of course, such lists can be modified and adapted to your own needs in case they do not fit perfectly.

## 13.5 To Go Further...

Text Mining and Natural Language Processing are topics that have been (and are still being) studied for a very long time. Recently, it has made a lot of progress thanks to the advances in technology and has gain even more interest with the abundance of text through social media, websites, blogs, etc. It is hence no surprise that a lot of machine learning models use text data (topic modelling, classification of emails to spam, etc.). Even current handy additions to simplify our life are based on text analysis (e.g. suggestions in emails, translation, etc.)

In the case you would want to go further on this topic, we strongly recommend the following books:

- *Text Mining with R*
- *Supervised Machine Learning for Text Analysis in R*
- *Textual Data Science with R*
- *R for Data Science* (through the introduction to web-scrapping etc.)

It is out of the scope of this book to develop this topic further, yet it is difficult to ignore the emergence of Generative AI (e.g. chatGPT) in the context of textual data. For example, it can be very helpful to automatically summarize comments from consumers (although it is outside the R framework).

# 14

# *Dashboards*

Since preparing the data and analyzing them is only a part of the story, we tackled in Chapter 6 how to generate a report from your analysis (and briefly how to analyze your data within your report) while Chapter 11 discussed points to consider to be as impactful in your communication as possible. For the latter, many formats on how to deliver and present your results were suggested, including the use of interactivity through dashboards. Since it is possible to build dashboard in R, we had to include a section that would introduce you to such solution. So embrace it, integrate it to your toolbox, and soon it will be your turn to shine during presentations!

## 14.1 Objectives

We have certainly been all through situations in which we spent a lot of time analyzing data for our study, built our report and our story, and spent time in perfecting our presentation. But when come the day of the presentation to your manager and colleagues, we get questions such as: *What would happen if we split the results between users/non users or between gender for instance?* In case you haven't been prepared for this question and didn't run the analysis up-front, you probably answered something like: *Let me rerun some analyses and I'll update you on that!*

Now imagine that you are in a similar situation, except that when such question arises, you have a tool that can answer live their questions using the data and the same analysis. Wouldn't that bring your discussion to another level? Even better: Imagine you can share this tool with your colleagues for them to use it and answer their own questions or use it for their own projects, even if they are not good at coding (nor even slightly familiar with R)?

Simply said, this is one of the roles of dashboards, as it brings interactivity to results (tables, graphs) by rerunning and updating them when data, options, and/or parameters are being altered.

DOI: 10.1201/9781003028611-14

The goal of this section is to build such dashboard using R and the {`shiny`} package.

## 14.2 Introduction to Shiny through an Example

### 14.2.1 What Is a Shiny Application?

In case you have already been through the visualization Chapter 5, you've already been briefly introduced to some sort of dashboard in R through the {`esquisse`} package. In this section, the {`shiny`} package is used to build such dashboard.

{`shiny`} is an R package that allows you to directly create from R interactive web applications. Its goal is to *convert* your R code into an application that can be accessible and used by anyone through their web browser, without having to be familiar with R.

This procedure is made available as {`shiny`} uses some carefully curated set of user interface (UI) functions that generate the HTML, CSS, and JavaScript code needed for most common tasks. In most cases, further knowledge of these languages is not required... unless you want to push your application further. Additionally, it introduces a new way of programming called *reactive programming*, which tracks automatically dependencies between code: When a change in an input is detected, any code that is affected by this change will automatically be updated.

### 14.2.2 Starting with Shiny

To create your very first shiny application, you can click on R studio in the *new page* icon and select *Shiny Web App...* Once you filled in the relevant information (name, author), you can then decide whether you want to create one unique file (`app.R`) or multiple files (`ui.R` and `server.R`).

Both solutions are equivalent and work the same: In both cases, a `ui()` and a `server()` function are generated. Due to better readability and to ease its maintenance over time, we recommend to use the single file for short applications and to use multiple files for larger applications (larger meaning with more code lines).

For our short application, it seems more convenient to use a single file.

### 14.2.3 Illustration

Let's consider a simple application in which we would import a data set that contains sensory data, as collected from a trained panel. In this example, the data set structure follows the one in *biscuits_sensory_profile.xlsx*.

For simplification purposes, the code developed for this application requires that the data set contains one column called *Judge* (containing the panelist information), one column called *Product* (containing the product information), and all the other columns being quantitative (corresponding to the sensory attributes).

The goal of the application is then to compute from this raw data set the sensory profiles of the products and to display them on screen. Furthermore, we also represent these sensory profiles graphically in a spider plot or in a circular bar plot. Since the main goal of shiny application is in its interactivity, the user should have the opportunity to remove/add/reorder the attributes on the table and plots and to hide/show products to increase visibility.

Once the graphics match our needs, we also propose to download it as a *.png* file to integrate it in our report.

From a learning perspective, this application introduces you specifically to:

- Importing an external file to the application;
- Create options that are both independent (type of graph to produce) and dependent on the file imported (list of attributes and products);
- Run some analyses (compute the means) and display the results as a table and as a plot;
- Export the graph to your computer as a png file.

---

The code used in this chapter in Section "User Interface" can be found in `app.R`. In the next sections, pieces of code are shown for explanation, and should/can not be run on their own. Instead, the entire application should be run.

---

### User Interface

The user interface (UI) is the part of the application that controls what the user sees and controls. In our example, the UI is separated into two parts:

- The left panel contains the options that the user can manually change;
- The right (or main) panel contains the outputs.

In the *app.R* file, this information is stored in the `ui()` function, and the two panels can be found in `sidebarPanel()` and in `mainPanel()`, respectively.

In `sidebarPanel()`, all the options are set up. These options include `fileInput()` for importing the data set or `radioButtons()` to control the type of plot to generate. A large list of options exist including `numericInput()`, `sliderInput()`, `textInput()`, `passwordInput()`, `dateInput()`, `selectInput()`, `checkboxInput()`, etc. Note that this library of options can be extended by adding `checkboxGroupInput()` from the {`shinyjs`} package.[1]

For most of these options, setting them up is quite straightforward, especially when (say) the range of values is already known beforehand (e.g. p-value ranging from 0 to 1, with default value at 0.05). However, in some cases, the option of interest cannot be defined on the UI side since they depend on the data itself (e.g. the product or attribute selection in our example). In such situations, these options are created on the server side and are retrieved on the UI side through `uiOutput()`.

On `mainPanel()`, `tabsetPanel()` and `tabPanel()` control the design of the output section. In our example, two tabs (one for the table and one for the graph) are created, although they could have been printed together on one page.

In our simple example, the `mainPanel()` is only used to export results computed on the server side. Depending on the type of output generated, the correct function used to retrieve the results should be used:

- For tables, `tableOutput()` is used to retrieve the table generated with `renderTable()`;
- For graphics, `plotOutput()` is used to retrieve the plot generated with `renderPlot()`;
- For elements to download, `downloadButton()` is used to retrieve the element (here a plot, but could be an Excel or PowerPoint file for instance) generated with `downloadHandler()`.

---

Note the pattern in the namings of the complementary functions: The `xxxOutput()` function (UI side) is used to retrieve the output generated (server side) by the corresponding `renderXxx()` function. This also applies with `uiOutput()` and `renderUI()`.

---

### Server

The server side of the application is where all the computations are being performed, including the construction of tables, figures, etc.

Since the options defined on the UI side should affect the computations performed (e.g. our decision on the type of plot to design should affect the plot generated), we need to communicate these decisions to the server and use them. On the server side, any information (or option) passed to the server side

[1] To use {`shinyjs`}, you need to load the library and add `useShinyjs()`, at the start of your `ui()` code.

is done through `input$name_option`. In our previous example regarding the type of graph to generate, this is shown as `input$plottype`, as defined by:

```
radioButtons("plottype", "Type of Plot to Draw:",
             choices=c("Spider Plot"="line", "Circular Barplot"="bar"),
             selected="line", inline=TRUE)
```

In this case, if the user selects *Spider Plot* (resp. *Circular Barplot*), `input$plottype` will take the value *line* (resp. *bar*).

In contrast, any information that is being build on the server side and that should be passed on the UI part of the application can either be done via the `xxxOutput()`/`renderXxx()` combination presented before (useful for showing results), including the `renderUI()`/`uiOutput()` combination (useful for options that are *server-dependent*).

Following a similar communication system than the one from UI to server, the part generated on the server side is stored as `output$name_option` (defined as `renderUI()`) and is captured on the UI side using `uiOutput ("name_option")`.

In our example, the latter combination is used for the two options that require reading the data set first, namely the selection of attributes and the selection of products.

```
# server side:
  output$attribute <- renderUI({

    req(mydata())

    items <- mydata() %>%
      pull(Attribute) %>%
      as.character() %>%
      unique()

    selectInput("attribute",
                "Select the Attributes (in order) ",
                choices=items, selected=items,
                multiple=TRUE)
  })

  output$product <- renderUI({

    req(mydata())
    items <- mydata() %>%
      pull(Product) %>%
      unique()
```

```
        checkboxGroupInput("product",
                           "Select the Products to Display:",
                           choices=items, selected=items)
  })

# UI side:
  uiOutput("attribute")
  uiOutput("product")
```

Lastly, we have elements that are only relevant on the server side, namely the computation themselves. In our example, these are results of a function called `reactive()`.

Reactivity (and its corresponding `reactive()` function) is a great *lazy* feature of {`shiny`} that was designed so that the computations are only performed when necessary, that is, when changes in an input affect computations. This laziness is of great power since only computations that are strictly needed are being performed, thereby increasing speed by limiting the computation power required to its minimum.

Let's break this down in a simple example:

```
  mydata <- reactive({
    req(input$datafile)
    data <- readxl::read_xlsx(input$datafile$datapath, sheet=1) %>%
      pivot_longer(-c(Judge, Product),
                   names_to="Attribute", values_to="Score") %>%
      mutate(Attribute = fct_inorder(Attribute),
             Score = as.numeric(Score)) %>%
      group_by(Product, Attribute) %>%
      summarize(Score = mean(Score)) %>%
      ungroup()
    return(data)
  })
```

In this section, the file selected by the user is read through the `fileInput()` option called `datafile` on the UI side. The path of the file is stored in the object called `datapath`, meaning that to access this file, we need to read `input$datafile$datapath`.

Once read (here using {`readxl`}), some small transformations to the data are performed before saving its final version in an object called `mydata`. Since this block of code only depends on `input$datafile` (UI side), this part is no longer used unless `datafile` is being updated or changed.

For the computation of the means, the same procedure applies as well:

```
mymean <- reactive({

  req(mydata(), input$attribute, input$product)
  mymean <- mydata() %>%
    mutate(across(c("Product", "Attribute"), as.character)) %>%
    filter(Attribute %in% input$attribute) %>%
    mutate(Product = factor(Product, input$product),
           Attribute = factor(Attribute, input$attribute),
           Score = format(round(Score, 2), nsmall=2)) %>%
    pivot_wider(names_from=Product, values_from=Score)

  return(mymean)
})
```

For this `reactive()` block, `mymean` depends on `mydata`, `input$attribute`, and `input$product`. This means that if `datafile` (read `mydata`), `input$attribute` and/or `input$product` change, the computations rerun and `mymean` is getting updated.

For small and simple examples like ours, this domain of reactivity may be sufficient, and would be sufficient in many cases. There are, however, some few points that require a bit more explanations.

First, we advise that you use `reactive()` as much as possible: In our example, we could have created the code to build the graph within `renderPlot()`. However, this way of coding is not efficient since it would always be updated, even when it is not necessary. For small examples such as the one proposed here, this may not make much difference, but for larger applications it would have a larger impact. This is why we prefer to create the graphs in a `reactive()` instance, and simply retrieve it for display.

Second, and as you may have seen already, the output of a `reactive()` section can be reused in other sections. This means that just like in *regular coding*, you can save elements in R object that you can reuse later (e.g. `mydata`, `mymean`, or `myplot`). However, these elements act like functions, meaning that if you want to call them, you should do it as `mydata()` for instance. More generally, let's imagine that `mydata` is a list with two elements (say `mydata$element1` and `mydata$element2`), we would retrieve `element1` as `mydata()$element1`.

Third, let's introduce the function `req()` that is used at the start of almost every block of code on the server side. To do so, let's take the example of `output$attribute` which starts with `req(mydata())`. The `req()` functions aims at requiring the object mentioned (here `mydata()`) before running: if `mydata()` doesn't exist, then `output$attribute` is set as `NULL`. This small line of code comes handy as it avoids returning design errors: How to extract a list of attributes from data that do not exist yet?

Finally, the application that we are developing here is *over-reactive* as every change we do will create update results. To highlight this, just remove some attributes from the list and you'll see the mean table or graphic being updated. In our small example, this is not too problematic since the application runs fast, but in other instances in which more computation is required, you may not want to wait that each little change done is being processed. To over come this, you can replace `reactive()` by `eventReactive()` combined with a button (e.g. *Run* or *Apply changes*) that only trigger changes once pressed. This means that changes are only performed on a user action.

### 14.2.4 Deploying the Application

To run the application, three options exist (within RStudio):

- Push the *Run app* button on the task bar of your script (a menu allows you to run the app in the Viewer window, or as a new window).
- Type directly `shiny::runApp('code')` in R.
- Use the shortcut *CTRL+SHIFT+ENTER* (on windows).

In this case, your computer (and RStudio) will use a virtual local server to build the application.

Unfortunately, this solution is not sufficient in case you really want to share it with colleagues. In such case, you need to publish your app by installing it on a server. Ideally, you have access to a server that can host such application (check with your IT department). If not, you can always create an account on https://www.shinyapps.io/admin/ and install it there. Note that there is a free registration option, which comes with limitations (number of applications to install). Also, before using their services, make sure that their conditions (privacy, protection of the data, etc.) suit you and your company policy.

## 14.3 To Go Further...

Quickly, {`shiny`} became popular and many researchers developed additional packages and tools to further enhance it. Without being exhaustive, here are a few tools, widgets, and packages that we often use as they provide interesting additional features. But don't hesitate to look online for features that answer your needs!

### 14.3.1 Personalizing and Tuning Your Application

If you start building many applications in {shiny}, you might get tired of its default layout. Fortunately, the overall appearance of applications can be modified easily, especially if you have notion in other programming languages such as CSS and HTML. If not, no worries, there are alternative solutions for you, including the {bslib} package. To change the layout, simply load the library and add the following piece of code at the start of your application (here for using the theme *darkly*):

```
fluidPage(
  theme = bslib::bs_theme(bootswatch = "darkly")
)
```

{bslib} also allows you creating your own theme that matches your company style. So do not hesitate to take some time to build it once and to apply it to all your applications.

Besides changing the overall theme of your applications, there are certain details that can make your overall application more appealing and easier to use. In our short example, you may have noticed that each option and each output is a separate line. This is sufficient here since the number of options and outputs are very limited. However, as the application grows in size, this solution is no longer sustainable as you do not want the users to scroll down a long way to adjust all the options.

Instead, you can combine different options (or outputs) on one line. To do so, {shiny} works by defining each panel as a table with as many rows as needed and 12 columns. When not specified, the whole width (i.e. the 12 columns) of the screen is used, but this could be controlled through the column() function.

In our example, we could have positioned the product selection and the attribute selection on the same line using:

```
fixedRow(
  column(6, uiOutput("product")),
  column(6, uiOutput("attribute"))
)
```

### 14.3.2 Upgrading Tables

The default table built in {shiny} using the tableOutput()/renderTable() combination is very handy, yet limited in its layout. Fortunately, many additional packages that create HTML tables have been developed to provide alternative solution to build more flexible tables.

In particular, we recommend the use of the {DT} and {rhandsontable} packages as they are very simple to use and yet provides a large variety of powerful design options. Just to name a few, it allows:

- cells or text formatting (e.g. coloring, rounding, adding currencies, or other units, etc.);
- merge rows or columns;
- add search/filter fields to columns of the table;
- provide interactivity, that for instance can be connected to graphics;
- include graphics within cells;
- allows manual editing, giving the user the chance to fill in or modify some cells.

To build such table, you will need to load the corresponding packages. For {DT} tables, you can generate and retrieve them using the complementary functions `renderDataTable()` and `dataTableOutput()`, or its concise forms `renderDT()` and `DTOutput()`.[2] For {rhandontable} tables, you can generate and retrieve them using `renderRHandsontable()` and `rHandsontable Output()`.

For more information and examples, please look at https://rstudio.github.io/DT/ and https://jrowen.github.io/rhandsontable/.

### 14.3.3 Building Dashboard

The example used here illustrates the power of {shiny}. However, it is limited to our own data set, meaning that it is study specific. What if we would want to create a dashboard that is connected to a database, for instance, and that updates its results as soon as new data are being collected?

This is of course the next step and {shiny} can handle this thanks to the {shinydashboard} package.

In its principle, {shinydashboard} works in a similar way to {shiny} itself, except for the structure of the UI. For {shinydashboard}, the UI contains three sections as shown below (the example below generates an empty dashboard):

```
library(shiny)
library(shinydashboard)
```

[2] The table should be generated using `datatable()`.

```
ui <- dashboardPage(
  dashboardHeader(),
  dashboardSidebar(),
  dashboardBody()
)

server <- function(input, output) { }

shinyApp(ui, server)
```

It is then your task to fill in the dashboard by, for instance, connecting to your data source (e.g. a simple data set, a database, etc.) and to fill in your different tabs with the interactivity and outputs of interest.

For more information, including a comprehensive *Get started* section, please visit https://rstudio.github.io/shinydashboard/

### 14.3.4 Interactive Graphics

Through its way of working, {`shiny`} creates some interactivity to graphics by updating it when changing some options. This is hence done by replacing a static graph by another static graph.

However, R provides other options that creates interactive graphs directly. This can be done thanks to the {`plotly`} library.

{`plotly`] is an alternative library to {`ggplot2`} that can be used to build R graphics within the R environment: It is not specific to {`shiny`} and can be used outside Shiny applications. To build {`plotly`} visualizations, you can build it directly from scratch using the `plot_ly()` function. It is out of the scope of this book to develop further how {`plotly`} works, mainly because we made the decision to explain in details how {`ggplot2`} works. And fortunately, {`plotly`} provides an easy solution to convert automatically {`ggplot2`} graph to {`plotly`} thanks to the `ggplotly()` function.

For more information, please visit https://plotly.com/r/

### 14.3.5 Interactive Documents

Ultimately, {`shiny`} can also be combined to other tools such as {`rmarkdown`} to build interactive tutorials, teaching material, etc. This is done by integrating the interactivity of {`shiny`} to propose options and reactive outputs into a text editor through {`rmarkdown`}.

To integrate {`shiny`} in your {`rmarkdown`}, simply add `runtime: shiny` in the YAML metadata of your R markdown document.

### 14.3.6 Documentation and Books

Thanks to its way of working and its numerous extensions, there is (almost) no limit to applications you can build (except maybe your imagination?). For inspiration, and to get a better idea of the powerful applications that you can build, have a look at the gallery on the official shiny webpage: https://shiny.rstudio.com/gallery/

In this section, we just introduced you to the main functions available in {`shiny`}, but if you want to go further, there is a whole world for you to explore. Of course, a lot of material is available online, and we would not know where to start to guide you. However, we strongly recommend you to start with the book from Hadley Wickham entitled *Mastering Shiny* (https://mastering-shiny.org/) as it is comprehensive and will give you the kick start that you need... and more.

# 15

# *Conclusion and Next Steps*

Congratulations, you've reached the end of this book!

We hope we have motivated you to continue your amazing journey into the emerging field of computational sensory science. To give you a little hand, we are listing here some other resources we would recommend and a summary of the main useful packages for sensory and consumer data analysis/visualization, including the ones we used throughout this book.

## 15.1 Other Recommended Resources

- *R for Data Science* (2nd edition) by Hadley Wickham, Mine Çetinkaya-Rundel, and Garrett Grolemund https://r4ds.hadley.nz/
- *Analyzing Sensory Data with R* by Sebastien Le and Thierry Worch
- *Rapid Sensory Profiling Techniques (2nd Edition)* by Julien Delarue and J. Ben Lawlor
- *Practical Guide to Cluster Analysis in R* by Alboukadel Kassambara
- *Practical Guide to Principal Component Methods in R* by Alboukadel Kassambara
- *Multiple Factor Analysis by Example Using R* by Jerome Pages
- *Using the flextable R package* by David Gohel (https://ardata-fr.github.io/flextable-book/index.html)
- *Tidy Modelling with R* by Max Kuhn and Julia Silge (https://www.tmwr.org/)
- *Hands-On Machine Learning with R* by Brad Boehmke and Brandon Greenwell (https://bradleyboehmke.github.io/HOML/)
- *Introduction to Statistical and Machine Learning Methods for Data Science* by Carlos Andre Reis Pinheiro and Mike Patetta
- *Supervised Machine Learning for Text Analysis in R* by Emil Hvitfeldt and Julia Sigle

DOI: 10.1201/9781003028611-15

- *R Graphics Cookbook: Practical Recipes for Visualizing Data* by Winston Chang
- *Text Mining with R: A Tidy Approach* by David Robinson and Julia Silge
- *Textual Data Science with R* by Mnica Bcue-Bertaut
- *Design and Analysis of Experiments with R* by John Lawson
- *Mastering Shiny* by Hadley Wickham (https://mastering-shiny.org/)

Some interesting books related to storytelling, graphical design, and data visualization:

- *Storytelling with Data* by Cole Nussbaumer Knaflic
- *Beyond Bullet Points* by Cliff Atkinson
- *Once Upon an Innovation* by Jean Storlie and Mimi Sherlock
- *Show me the Number: Designing Table and Graphs to Enlighten* by Stephen Few

More generally, the emergence of Generative AI cannot and should not be ignored, as it can be a great digital assistant in many situations. In the particular context of this book, it can be of great support for coding, as it can generate some for you (e.g. you can find videos online on how to build Shiny Applications using ChatGPT). It can also be used to help you find and fix errors in your script. However, be aware that the AI-generated code might be far from perfect, and your expertise and understanding of coding language is extremely valuable to ensure that what is done is correct. Be also aware of any other of its limitations including privacy before using it.

## 15.2 Useful R Packages

- `ClustBlock`: hierarchical and partitioning algorithms of blocks of variables. Includes functions for clustering subjects and multivariate analysis of multiblock sensory data, such as Check-All-that-Apply (CLUSCATA) or Free Sorting (CLUSTATIS).
- `ClustVarLV`: functions for the clustering of variables around Latent Variables, for two-way or three-way data.
- `corrplot`: provides a visual exploratory tool on correlation matrix that supports automatic variable reordering to help detect hidden patterns among variables.

- `DistatisR`: implements three-way multidimensional scaling. DiSTATIS is used to analyze multiple distance matrices collected on the same set of observation (e.g. by free sorting).
- `FactoExtra`: makes easy to extract and visualize the output of exploratory multivariate data analyses, including Principal Component Analysis (PCA), Correspondence Analysis (CA), Multiple Correspondence Analysis (MCA), Multiple Factor Analysis (MFA), Hierarchical Clustering (HCKUST), and partioning Clustering (e.g. k-means, PAM,CLARA, etc.)
- `FactoMineR`: dedicated to multivariate Exploratory Data Analysis including Principal Components Analysis (PCA), Correspondence analysis (CA), Multiple Correspondence Analysis (MCA), and clustering.
- `rstatix`: a pipe-friendly framework, coherent with the "tidyverse" design philosophy, for performing basic statistical tests.
- `sensmixed`: to analyze sensory and consumer data within mixed effects model framework.
- `SensoMineR`: dedicated to the statistical analysis of sensory data. It tackles the characterization of the products, panel performance assessment, links between sensory and instrumental data, consumer's preferences, napping evaluation, and optimal designs.
- `SensR`: for Thurstonian Models for sensory discrimination methods, including duotrio, tetrad, triangle, 2-AFC, 3-AFC, A-not A, same-different, 2-AC, and degree-of-difference. This package enables the calculation of d-primes, standard errors of d-primes, sample size and power computations, and comparisons of different d-primes.
- `stats`: contains functions for statistical calculations and random number generation. The analysis include ANOVA, posthoc tests, Clustering, Correlation, multivariate analysis, among many others.
- `tempR`: for analysis and visualization of data from temporal sensory methods, including temporal check-all-that-apply (TCATA) and temporal dominance of sensation.

# *Bibliography*

Ares, G. and Varela, P. (2017). Trained vs. consumer panels for analytical testing: Fueling a long lasting debate in the field. *Food Quality and Preference*, 61:79–86.

Atkinson, C. (2018). *Beyond bullet points: Using PowerPoint to Tell a Compelling Story That Gets Results.* Pearson Education, Inc., 4th edition.

Baardseth, P., Bjerke, F., Aaby, K., and Mielnik, M. (2005). A screening experiment to identify factors causing rancidity during meat loaf production. *European Food Research and Technology*, 221(5):653–661.

Bécue-Bertaut, M. (2019). *Textual Data Science with R.* Boca Raton, FL, Chapman & Hall/CRC Press.

Bécue-Bertaut, M., Álvarez-Esteban, R., and Pagès, J. (2008). Rating of products through scores and free-text assertions: Comparing and combining both. *Food Quality and Preference*, 19(1):122–134.

Ben Slama, M., Heyd, B., Danzart, M., and Ducauze, C. (1998). Plans d-optimaux: une stratégie de réduction du nombre de produits en cartographie des préférences. *Sciences des aliments*, 18(5):471–483.

Bleibaum, R. N., editor (2020). *Descriptive Analysis Testing for Sensory Evaluation.* ASTM International, 2nd edition.

Blundell, J., De Graaf, C., Hulshof, T., Jebb, S., Livingstone, B., Lluch, A., Mela, D., Salah, S., Schuring, E., Van Der Knaap, H., and Westerterp, M. (2010). Appetite control: Methodological aspects of the evaluation of foods. *Obesity Reviews*, 11(3):251–270.

Brockhoff, P.B. (2011). Sensometrics. In: Lovric, M. (ed.) *International Encyclopedia of Statistical Science.* Springer, Berlin, Heidelberg. pp. 1302–1305, https://doi.org/10.1007/978-3-642-04898-2_69

Bryan, J. (2018). Excuse me, do you have a moment to talk about version control? *The American Statistician*, 72(1):20–27.

Cadoret, M., Lê, S., and Pagès, J. (2009). A factorial approach for sorting task data (FAST). *Food Quality and Preference*, 20(6).410–417.

Cao, L. (2017). Data science: A comprehensive overview. *ACM Computing Surveys*, 50(3):1–42.

Cariou, V. and Wilderjans, T. F. (2018). Consumer segmentation in multi-attribute product evaluation by means of non-negatively constrained clv3w. *Food Quality and Preference*, 67:18–26.

Civille, G. V. and Carr, B. T. (2015). *Sensory evaluation techniques.* CRC Press.

Cleveland, W. S. (2001). Data science: An action plan for expanding the technical areas of the field of statistics. *International Statistical Review*, 69(1):21–26.

Dairou, V., Priez, A., Sieffermann, J. M., and Danzart, M. (2003). An original method to predict brake feel: A combination of design of experiments and sensory science. *SAE Transactions*, 112(0598):735–741.

Danzart, M. (1998). Quadratic model in preference mapping. *4th Sensometrics meeting*, Copenhagen, Denmark.

Danzart, M., Sieffermann, J. M., and Delarue, J. (2004). New developments in preference mapping techniques: Finding out a consumer optimal product, its sensory profile and the key sensory attributes. *7th Sensometrics Conference*, Davis, CA.

Davenport, T. H. and Patil, D. J. (2012). Data scientist. *Harvard Business Review*, 90(5):70–76.

Dean, A., Voss, D., and Draguljic, D. (2017). *Design and Analysis of Experiments.* Springer, 2nd edition.

Delarue, J. and Lawlor, J. B. (2022). *Rapid Sensory Profiling Techniques. Applications in New Product Development and Consumer Research.* Woodhead Publishing Ltd, Cambridge, UK, 2nd edition.

Delarue, J. and Sieffermann, J.-M. (2004). Sensory mapping using Flash profile. Comparison with a conventional descriptive method for the evaluation of the flavour of fruit dairy products. *Food Quality and Preference*, 15(4):383–392.

Ennis, D. M. (2016). *Thurstonian Models: Categorical Decision Making in the Presence of Noise.* Institute for Perception. Richmond, VA ISBN: 9780990644606

Fayyad, U., Piatetsky-Shapiro, G., and Smyth, P. (1996). From data mining to knowledge discovery in databases. *AI Magazine*, 17(3):37.

Feldmeyer, A. and Johnson, A. (2022). Using Twitter to model consumer perception and product development opportunities: A use case with Turmeric. *Food Quality and Preference*, 98:104499.

Few, S. (2012). *Show Me the Numbers: Designing Tables and Graphs to Enlighten.* Analytics Press.

Fisher, R. A. (1935). *The Design of Experiments.* Hafner Press, Macmillan Publishing, New York.

Franczak, B. C., Browne, R. P., McNicholas, P. D., and Findlay, C. J. (2015). Product selection for liking studies: The sensory informed design. *Food Quality and Preference*, 44:36–43.

Gacula, M. C. (2008). *Design and Analysis of Sensory Optimization.* Harvard Educational Review. Wiley, Trumbull, CT.

Galiñanes Plaza, A., Delarue, J., and Saulais, L. (2019). The pursuit of ecological validity through contextual methodologies. *Food Quality and Preference*, 73:226–247.

Hamilton, L. M. and Lahne, J. (2020). Fast and automated sensory analysis: Using natural language processing for descriptive lexicon development. *Food Quality and Preference*, 83:103926.

Husson, F., Le, S., and Pagès, J. (2017). *Exploratory Multivariate Analysis by Example Using R.* Chapman and Hall/CRC.

Hvitfeldt, E. and Silge, J. (2021). *Supervised Machine Learning for Text Analysis in R.* Chapman and Hall/CRC, New York.

ISO11035 (1995). Sensory analysis-identification and selection of descriptors for establishing a sensory profile by a multidimensional approach. Standard.

ISO8586 (2012). Sensory analysis–general guidelines for the selection, training and monitoring of selected assessors and expert sensory assessors. Standard.

Jaeger, S. R., Hunter, D. C., Kam, K., Beresford, M. K., Jin, D., Paisley, A. G., Chheang, S. L., Roigard, C. M., and Ares, G. (2015). The concurrent use of jar and cata questions in hedonic scaling is unlikely to cause hedonic bias, but may increase product discrimination. *Food Quality and Preference*, 44:70–74.

Kahneman, D. and Tversky, A. (2000). *Choices, Values and Frames.* Cambridge University Press and the Russell Sage Foundation, Cambridge.

Kassambara, A. (2017a). *Practical Guide to Cluster Analysis in R: Unsupervised Machine Learning.* CreateSpace Independent Publishing Platform, 1st edition.

Kassambara, A. (2017b). *Practical Guide to Principal Component Methods in R: PCA, M(CA), FAMD, MFA, HCPC, factoextra.* Multivariate Analysis. Sthda.com.

Knaflic, C. N. (2015). *Storytelling with Data.* John Wiley & Sons, Hoboken, NJ.

Köster, E. (2003). The psychology of food choice: Some often encountered fallacies. *Food Quality and Preference*, 14:359–373.

Köster, E., Couronne, T., Léon, F., Lévy, C., and Marcelino, A. S. (2003). Repeatability in hedonic sensory measurement: A conceptual exploration. *Food Quality and Preference*, 14:165–176.

Lawless, H. T. and Heymann, H. (2010). *Sensory Evaluation of Food: Principles and Practices.* Food Science Text Series. Springer New York, 2nd edition.

Lawson, J. (2014). *Design and Analysis of Experiments with R.* Chapman and Hall/CRC, 1st edition.

Lawson, J. and Willden, C. (2016). Mixture experiments in R using mixexp. *Journal of Statistical Software*, 72(2):1–20.

Lê, S. and Worch, T. (2018). *Analyzing Sensory Data with R.* Chapman and Hall/CRC.

Lee, H.-S. and O'Mahony, M. (2004). Sensory difference testing: Thurstonian models. *Food Science and Biotechnology*, 13(6):841–847.

Liu, B. (2015). *Sentiment Analysis.* Cambridge University Press.

Llobell, F., Cariou, V., Vigneau, E., Labenne, A., and Qannari, E. M. (2020). Analysis and clustering of multiblock datasets by means of the statis and clustatis methods. Application to sensometrics. *Food Quality and Preference*, 79:103520.

Llobell, F., Vigneau, E., and Qannari, E. M. (2019). Clustering datasets by means of clustatis with identification of atypical datasets. Application to sensometrics. *Food Quality and Preference*, 75:97–104.

Luc, A., L, S., Philippe, M., Qannari, E. M., and Vigneau, E. (2022a). Free jar experiment: Data analysis and comparison with jar task. *Food Quality and Preference*, 98:104453.

Luc, A., Lê, S., and Philippe, M. (2020). Nudging consumers for relevant data using Free JAR profiling: An application to product development. *Food Quality and Preference*, 79:103751.

Luc, A., Lê, S., Philippe, M., Mostafa Qannari, E., and Vigneau, E. (2022b). A machine learning approach for analyzing Free JAR data. *Food Quality and Preference*, 99:104581.

MacFie, H. (2007). Preference mapping and food product development. In MacFie, H., editor, *Consumer-led Food Product Development*, Woodhead Publishing Series in Food Science, Technology and Nutrition, pp. 551–592. Woodhead Publishing Ltd, Cambridge.

Macfie, H. J., Bratchell, N., Greenhoff, K., and Vallis, L. V. (1989). Designs to balance the effect of order of presentation and first-order carry-over effects in hall tests. *Journal of Sensory Studies*, 4(2):129–148.

Mahieu, B., Visalli, M., and Schlich, P. (2022). Identifying drivers of liking and characterizing the ideal product thanks to Free-Comment. *Food Quality and Preference*, 96:104389.

Mahieu, B., Visalli, M., Thomas, A., and Schlich, P. (2020). Free-Comment outperformed check-all-that-apply in the sensory characterisation of wines with consumers at home. *Food Quality and Preference*, 84:103937.

Mao, M. and Danzart, M. (2007). How to select the best subset of factors maximizing the quality of multi-response optimization. *Quality Engineering*, 20(1):63–74.

Meiselman, H., editor (2019). *Contet. The Effects of Environment on Product Design and Evaluation*. Woodhead Publishing.

Meullenet, J.-F., Xiong, R., and Findlay, C. J. (2008). *Multivariate and Probabilistic Analyses of Sensory Science Problems*. John Wiley & Sons.

Modi, V. K. and Prakash, M. (2008). Quick and reliable screening of compatible ingredients for the formulation of extended meat cubes using PlackettBurman design. *LWT – Food Science and Technology*, 41(5):878–882.

Mohammad, S. M. and Turney, P. D. (2013). Crowdsourcing a word-emotion association lexicon. *Computational Intelligence*, 29(3):436–465.

Murrell, P. (2011). *R Graphics*. CRC Press, 2nd edition.

Naes, T. and Nyvold, T. E. (2004). Creative design–An efficient tool for product development. *Food Quality and Preference*, 15(2):97–104.

Naur, P. (1974). *Concise Survey of Computer Methods*. Petrocelli Books.

O'Mahony, M. (1986). *Sensory Evaluation of Food: Statistical Methods and Procedures*. Routledge.

O'Mahony, M. and Rousseau, B. (2003). Discrimination testing: A few ideas, old and new. *Food Quality and Preference*, 14(2):157–164.

Peng, R. D. (2011). Reproducible research in computational science. *Science*, 334(6060):1226–1227.

Perrin, L., Symoneaux, R., Maître, I., Asselin, C., Jourjon, F., and Pagès, J. (2008). Comparison of three sensory methods for use with the Napping® procedure: Case of ten wines from Loire valley. *Food Quality and Preference*, 19(1):1–11.

Peryam, D. R. and Pilgrim, F. J. (1957). Hedonic scale method of measuring food preferences. *Food Technology*, 11:9–14.

Petiot, J. F. (2022). A Genetic Approach for the Interactive Design of Sounds: Application to Electric Vehicles. In: *Nonfood Sensory Practices*, edited by Anne-Marie Pensé-Lhéritier, Irène Bacle, and Julien Delarue, 251–271. Woodhead Publishing Series in Food Science, Technology and Nutrition. Woodhead Publishing, 2022. https://doi.org/10.1016/B978-0-12-821939-3.00017-8.

Phetxumphou, K., Cox, A. N., and Lahne, J. (2020). Development and characterization of a check-all-that-apply (CATA) Lexicon for Virginia hard (alcoholic) ciders. *Journal of the American Society of Brewing Chemists*, 78(4):299–307.

Pineau, N., Girardi, A., Lacoste Gregorutti, C., Fillion, L., and Labbe, D. (2022). Comparison of rata, cata, sorting and napping as rapid alternatives to sensory profiling in a food industry environment. *Food Research International*, 158:111467.

Pineau, N., Moser, M., Rawyler, F., Lepage, M., Antille, N., and Rytz, A. (2019). Design of experiment with sensory data: A pragmatic data analysis approach. *Journal of Sensory Studies*, 34(2):e12489.

Piqueras-fiszman, B. (2015). Open-ended questions in sensory testing practice. In *Rapid Sensory Profiling Techniques and Related Methods: Applications in New Product Development and Consumer Research*, pp. 247–267. Woodhead Publishing Ltd.

Popper, R., Rosenstock, W., Schraidt, M., and Kroll, B. (2004). The effect of attribute questions on overall liking ratings. *Food Quality and Preference*, 15(7):853–858. Fifth Rose Marie Pangborn Sensory Science Symposium.

Porter, M. (1980). An algorithm for suffix stripping. *Program*, 14(3):130–137.

Prescott, J., Lee, S. M., and Kim, K.-O. (2011). Analytic approaches to evaluation modify hedonic responses. *Food Quality and Preference*, 22(4):391–393.

Qannari, E. M. (2017). Sensometrics approaches in sensory and consumer research. *Current Opinion in Food Science*, 15:8–13.

R Core Team (2022). *R: A Language and Environment for Statistical Computing*. R Foundation for Statistical Computing, Vienna, Austria.

Rasch, D., Pilz, J., Verdooren, R., and Gebhardt, A. (2011). *Optimal Experimental Design with R*. CRC Press, Taylor & Francis.

Rivière, P., Monrozier, R., Rogeaux, M., Pagès, J., and Saporta, G. (2006). Adaptive preference target: Contribution of kanos model of satisfaction for an optimized preference analysis using a sequential consumer test. *Food Quality and Preference*, 17(7):572–581.

Rothman, L. and Parker, M. (2009). *Just-About-Right (JAR) Scales*. West Conshohocken, PA: ASTM International.

Rytz, A., Moser, M., Lepage, M., Mokdad, C., Perrot, M., Antille, N., and Pineau, N. (2017). Using fractional factorial designs with mixture constraints to improve nutritional value and sensory properties of processed food. *Food Quality and Preference*, 58:71–75.

Schlich, P. (1993). *Contribution à la sensométrie*. PhD thesis, Université Paris-Sud.

Schlich, P. and McEwan, J. (1992). Cartographie des préférences-un outil statistique pour l'industrie agro-alimentaire. *Sciences des aliments*, 12:339–355.

Silge, J. and Robinson, D. (2017). *Text Mining with R: A Tidy Approach*. O'Reilly Media, Inc.

Stone, H., Bleibaum, R. N., and Thomas, H. A., editors (2020). *Sensory Evaluation Practices*. Academic Press, 5th edition.

Storlie, J. and Sherlock, M. (2020). *Once Upon an Innovation: Business Storytelling Techniques for Creative Problem Solving*. Beaver's Pond Press.

Stubbs, R. J., Hughes, D. A., Johnstone, A. M., Rowley, E., Reid, C., Elia, M., Stratton, R., Delargy, H., King, N., and Blundell, J. (2000). The use of visual analogue scales to assess motivation to eat in human subjects: A review of their reliability and validity with an evaluation of new hand-held computerized systems for temporal tracking of appetite ratings. *British Journal of Nutrition*, 84(4):405–415.

Stunkard, A. J. and Messick, S. (1985). The three-factor eating questionnaire to measure dietary restraint, disinhibition and hunger. *Journal of Psychosomatic Research*, 29(1):71–83.

ten Kleij, F. and Musters, P. A. (2003). Text analysis of open-ended survey responses: A complementary method to preference mapping. *Food Quality and Preference*, 14(1):43–52.

Tukey, J. W. (1962). The future of data analysis. *The Annals of Mathematical Statistics*, 33(1):1–67.

Tukey, J. W. (1977). *Exploratory data analysis*, volume 2. Reading, MA.

Varela, P. and Ares, G. (2012). Sensory profiling, the blurred line between sensory and consumer science. A review of novel methods for product characterization. *Food Research International*, 48(2):893–908.

Vidal, L., Ares, G., Machín, L., and Jaeger, S. R. (2015). Using Twitter data for food-related consumer research: A case study on what people say when tweeting about different eating situations. *Food Quality and Preference*, 45:58–69.

Vigneau, E., Qannari, E., Navez, B., and Cottet, V. (2016). Segmentation of consumers in preference studies while setting aside atypical or irrelevant consumers. *Food Quality and Preference*, 47:54–63.

Visalli, M., Mahieu, B., Thomas, A., and Schlich, P. (2020). Automated sentiment analysis of Free-Comment: An indirect liking measurement? *Food Quality and Preference*, 82:103888.

Wakeling, I. N. and MacFie, H. J. (1995). Designing consumer trials balanced for first and higher orders of carry-over effect when only a subset of k samples from t may be tested. *Food Quality and Preference*, 6(4):299–308.

Wichchukit, S. and O'Mahony, M. (2015). The 9-point hedonic scale and hedonic ranking in food science: Some reappraisals and alternatives. *Journal of the Science of Food and Agriculture*, 95(11):2167–2178.

Wickham, H. (2014). Tidy data. *Journal of Statistical Software*, 59(1):1–23.

Wickham, H. and Grolemund, G. (2016). *R for Data Science: Import, Tidy, Transform, Visualize, and Model Data.* O'Reilly Media, Inc.

Wilderjans, T. F. and Cariou, V. (2016). Clv3w: A clustering around latent variables approach to detect panel disagreement in three-way conventional sensory profiling data. *Food Quality and Preference*, 47:45–53.

Williams, A. and Langron, S. (1984). The use of free-choice profiling for the evaluation of commercial ports. *Journal of the Science of Food and Agriculture*, 35(5):558–568.

Worch, T., Lê, S., Punter, P., and Pagès, J. (2013). Ideal profile method (ipm): The ins and outs. *Food Quality and Preference*, 28(1):45–59.

Wu, C. F. J. (1997). Statistics = Data Science? Lecture at the University of Michigan-Ann Arbor. Retrieved from: https://www2.isye.gatech.edu/~jeffwu/presentations/datascience.pdf

Yu, P., Low, M. Y., and Zhou, W. (2018). Design of experiments and regression modelling in food flavour and sensory analysis: A review. *Trends in Food Science & Technology*, 71:202–215.

# *Index*

Pages in *italics* refer to figures, pages in **bold** refer to tables.

**F**



**G**



**H**



**I**



**K**



**L**



**M**



**N**

## O



## P



## R



## S



## T

## U



## V



## W



## X



## Y